Advances in

Plant Pathology

Volume 3

Mathematical Modelling of Crop Disease

Advances in Plant Pathology

series edited by

D. S. Ingram
The Botany School
The University of Cambridge
Cambridge
England

and

P. H. Williams
Department of Plant Pathology
The University of Wisconsin
Madison, Wisconsin
USA

Volume 3

Mathematical Modelling of Crop Disease

edited by

Christopher A. Gilligan

Department of Applied Biology
The University of Cambridge
Cambridge
England

1985

ACADEMIC PRESS, INC.
Harcourt Brace Jovanovich, Publishers

London Orlando San Diego New York
Austin Montreal Sydney Tokyo Toronto

ACADEMIC PRESS INC. (LONDON) LTD.
24–28 Oval Road
LONDON NW1 7DX

United States Edition published by
ACADEMIC PRESS, INC.
Orlando, Florida 32887

ISBN 0–12–033703–7

PRINTED IN THE UNITED STATES OF AMERICA

85 86 87 88 9 8 7 6 5 4 3 2 1

Contents

4. Construction of Temporal Models: III. Disease Progress of Soil-Borne Pathogens
C. A. Gilligan

5. Construction of Dispersal Models
H. A. McCartney and B. D. L. Fitt

6. Genetic Modelling for Crop Disease Management
D. R. MacKenzie and M. H. Royer

7. Construction of Predictive Models: I. Forecasting Disease Development
W. E. Fry and G. R. Fohner

Contributors

Numbers in parentheses indicate the pages on which the authors' contributions begin.

D. J. BUTT (207), *Plant Pathology Department, East Malling Research Station, Maidstone ME19 6BJ, UK*

S. P. EISENSMITH (31), *Tropeninstitut, Abteilung Phytopathologie, Justus-Liebig-Universität, D-6300 Giessen, Federal Republic of Germany*

B. D. L. FITT (107), *Rothamsted Experimental Station, Harpenden, Hertfordshire AL5 2JQ, UK*

G. R. FOHNER[1] (161), *Department of Agricultural Economics, Cornell University, Ithaca, New York 14853, USA*

W. E. FRY (161), *Department of Plant Pathology, Cornell University, Ithaca, New York 14853, USA*

C. A. GILLIGAN (1, 67), *Department of Applied Biology, University of Cambridge, Cambridge CB2 3DX, UK*

B. HAU (31), *Tropeninstitut, Abteilung Phytopathologie, Justus-Liebig-Universität, D-6300 Giessen, Federal Republic of Germany*

M. J. JEGER (207), *Department of Plant Pathology and Microbiology, Texas A&M University, College Station, Texas 77843, USA*

J. KRANZ (31), *Tropeninstitut, Abteilung Phytopathologie, Justus-Liebig-Universität, D-6300 Giessen, Federal Republic of Germany*

[1] Present address: Goldsmith Seeds, Inc., Gilroy, California 95021, USA.

D. R. MACKENZIE (145), *Department of Plant Pathology and Crop Physiology, Louisiana State University, Baton Rouge, Louisiana 70803, USA*

H. A. MCCARTNEY (107), *Rothamsted Experimental Station, Harpenden, Hertfordshire AL5 2JQ, UK*

R. RABBINGE (231), *Department of Theoretical Production Ecology, Agricultural University, 6708 PD Wageningen, The Netherlands*

D. I. ROUSE (11), *Department of Plant Pathology, University of Wisconsin—Madison, Madison, Wisconsin 53706, USA*

M. H. ROYER (145), *Plant Disease Research Laboratory, USDA–ARS, Frederick, Maryland 21701, USA*

P. S. TENG (179), *Department of Plant Pathology, University of Minnesota, St. Paul, Minnesota 55108, USA*

J. C. ZADOKS (231), *Laboratory of Phytopathology, Agricultural University, 6709 PD Wageningen, The Netherlands*

Series Preface

Advances in Plant Pathology is a series which publishes extended review articles on topics of current or future interest and importance to plant pathologists.

Some volumes in the series, like Vols. 1 and 2, will not have any particular theme, but will contain a mixture of articles drawn from a range of subject areas within or related to the general field of plant pathology. Others will deal with a single important topic in depth, either as a series of short articles assembled by a specialist editor, like the present volume, or as a complete work by only one or two authors. Also, occasionally, a volume may be devoted to the proceedings of a symposium or conference. The contributors to all volumes are asked to concentrate on ideas and interpretations, to be controversial where appropriate, and to write for a wide audience which includes undergraduate and post-graduate students, research scientists and college teachers.

It is through its flexibility of style, format and content, its emphasis on ideas and interpretations, and its breadth of target audience that *Advances in Plant Pathology* is intended to differ from other review series. As the editors we are anxious to serve the needs and interests of all plant pathologists. We would therefore *welcome* comments, suggestions and ideas for future volumes. These should be sent to Dr. D. S. Ingram, *The Botany School, Downing Street, Cambridge CB2 3EA, England.*

Cambridge and Madison, 1984

D. S. Ingram
P. H. Williams

Preface

Interest in the application of mathematical models to the investigation of crop disease is increasing rapidly. This volume of *Advances in Plant Pathology* provides a synthesis of ideas on the current status and future development of mathematical models in major areas within botanical epidemiology. The optimism of this objective is not divorced from the realisation that the intrusion of mathematics into plant pathology is not universally welcomed. The book is addressed therefore not only to the devotees of mathematical modelling but also to those who are sceptical of the value of this approach to plant pathology.

The chapters are planned and arranged so that they constitute a theme: they progress from discussions of the construction of detailed models of specific aspects of crop disease to more applied aspects, firstly of forecasting disease or crop loss and secondly of implementing models to assist in the management of disease. The final chapter addresses the medium- to long-term perspective of the use of mathematical models for disease management.

The chapters are wide ranging with extensive reviews of the literature that extend beyond plant pathology to encompass related work in agronomy and plant physiology, entomology and population dynamics, genetics, statistics, mathematics, physics, economics and medical epidemiology. Nevertheless, in keeping with the intent of the series, the emphasis in this volume is on ideas rather than on intensive reviews. Alternative approaches to modelling are presented and discussed within and amongst chapters, wherein much consensus but also some dissension is to be found. I hope that the ideas presented in this volume will stimulate thought and much further work on the construction and successful implementation of models for crop disease.

Finally, and most importantly, I thank the contributors for their sustained efforts in making this publication possible.

Cambridge, 1984 Christopher A. Gilligan

1

INTRODUCTION

Christopher A. Gilligan

Department of Applied Biology, University of Cambridge, Cambridge, England

Mathematical modelling of crop disease is a rapidly expanding discipline within plant pathology. It has a relatively short history, compared with analogous disciplines, such as medical epidemiology and insect population dynamics, and is further distinguished by being largely the province of the biologist rather than of the mathematician. In spite of this, the emerging discipline is viewed with misgivings by many plant pathologists. A dichotomy is evident between modellers and non-modellers. Between the two groups a gulf of mutual incomprehension is threatened as the level of mathematical sophistication in epidemiological models increases (cf. Snow, 1964). Modellers themselves do not proceed unanimously: polarization of views is almost inevitable between protagonists of intrinsically different approaches to modelling, as, for example, between those who favour the use of complex simulators to comparatively simple growth equations to desribe the progress of disease. Communication is essential therefore not only between modellers and nonmodellers but also amongst modellers.

The first syntheses of ideas on the mathematical treatment of botanical epidemics were published just a quarter of a century ago by Van der Plank (1960, 1963) and Schrödter (1960). Prior to that time few papers containing substantial mathematical analysis of epidemiological problems had been published. Notable amongst these early papers were those of Gregory (1945, 1948), Rombakis (1947), Schrödter (1952a,b, 1954) and Waggoner (1952) on dispersal of propagules; Cochran (1936), Kleckowski (1950) and Van der Plank (1946, 1947, 1948, 1949) on the spatial pattern of disease; Large (1952) on disease progress, and Wilcoxon and McCallan (1939) and Dimond *et al.* (1941) on dose–response curves. Of these papers, however, only that of Gregory (1948), in which the multiple infection transformation was introduced, has been widely cited: the rest have received comparatively little attention until recently. The analyses of the temporal progress of disease by Van der Plank (1960, 1963) provided a stimulus for much further work. Following another synthesis of ideas (Kranz, 1974a), this aspect of epidemiology has developed rapidly (Rouse, Chapter 2, Hau *et al.*, Chapter 3). Mathematical modelling of the spatial progress of disease and of the dispersal of inoculum, however, has attracted comparatively less attention (Gregory, 1973; Aylor, 1978; McCartney and Fitt, Chapter 5). Concomitant with the expansion of mathematical modelling of epidemics has been increasing work on the measurement and estimation of levels of disease (Chester, 1950; Colwell, 1956; James, 1974; James and Teng, 1979) and also on the reappraisal of traditional methods of field experimentation (James and Shih, 1973; Jenkyn *et al.*, 1979).

Of the cognate sciences to botanical epidemiology, medical epidemiology, not surprisingly, has a long history. Bailey (1975) cites the work of Bernouille (1760) as the first use of a mathematical model to study the population dynamics of an infectious disease of man. Rapid developments in medical epidemiology occurred during the early decades of the twentieth century. Much of the early work was deterministic. However, following the publication of the first stochastic models by McKendrick (1926) and Greenwood (1931), the emphasis shifted from deterministic to stochastic models. Typically these take the form of compartmental models (Anderson, 1981) in which the probability of the flow of hosts between certain classes such as suscepts, infecteds and immunes is modelled, given conditions on the latent and infectious periods of the infectious agents. Little thought has yet been given in plant pathology to stochastic models other than the use of pseudo-random number generators in Monte Carlo simulation (Sall, 1980).

Mathematical work on the dynamics of animal populations including insects first flourished over half a century ago (Pearl and Reed, 1920; Thompson, 1924, 1930; Lotka, 1925, 1932; Yule, 1925; Volterra, 1926; Nicholson and Bailey, 1935; Gause, 1934). The classical work of this period greatly influenced subsequent work on animal populations (Pielou, 1977). Modelling of insect populations has remained part of the mainstream of ecological modelling (Hassell, 1976; May, 1981), thereby both influencing and contributing to developments in related fields. Thus, much attention has been given to models for the regulation of population size in artificial and semi-natural populations (Varley *et al.*, 1973; May, 1978) and hence to the behaviour of the population about the carrying capacity. More recent work by Anderson (1981), Anderson and May (1979) and May and Anderson (1979) has combined the approaches of population dynamics with those of medical epidemiology to model the population dynamics of infectious disease agents in natural communities.

Plant pathology, by contrast, has remained outside this coterie and mathematical modelling of crop disease has proceeded independently of that in other disciplines. Interest has centred on deterministic models for population growth and on models to forecast disease and crop loss. Relatively little attention has so far been given explicitly to regulation of populations of parasites (Jeger, 1982) and to the behavior of populations about the carrying capacity. In this respect, modelling of nematode populations is exceptional. Being a zoological rather than a botanical discipline, it reflects more closely developments in general ecological theory, such as the behaviour of populations about the carrying capacity (Seinhorst, 1965, 1970; Jones *et al.*, 1967; Jones and Perry, 1978).

I have used the term mathematical modelling freely without so far defining it. Mathematical modelling in botanical epidemiology involves the use of mathematics to formalise hypotheses concerning the behaviour of epidemics of plant disease. The symbolic logic inherent in the use of mathematics (Jeffers, 1978) furnishes a set of rules by which variables are constrained to behave. Examples of frequently used constraints are asymptotic functions to limit the maximum level a disease can attain and functions for negative feedback to slow the progress of disease as an epidemic progresses. The assumption of additivity of parameters in multiple regression is another commonly used mathematical constraint. Given these rules, mathematical models are constructed to satisfy a set of *a priori* assumptions proposed by the experimenter, on the basis of knowledge or intuition about the behaviour of the system. Implicit in this exercise is the desire to reduce a complex system to a series of simpler hypotheses that can be tested. A model, therefore, is not a replica of the system it serves to represent but some approximation to it (Kranz, 1974b).

Mathematical and statistical procedures provide a framework for the construction, fitting and testing of models, subject to specified assumptions. They do not in themselves, however, necessarily dictate the type of model nor the methodology of modelling to be followed. The objectives of modelling, together with the intuition and expertise of the investigator, also influence the choice.

Consider first the objectives of mathematical modelling of crop disease. Three broad objectives can be identified: description, prediction and explanation (Kranz and Royle, 1978). These are not mutually exclusive but represent instead the extremes of a three-dimensional spectrum of objectives. A model that explains a part of a system must inevitably have good predictive value. It does not follow, though, that a model that predicts a system well is necessarily explanatory, nor intuitively descriptive of the epidemic.

Descriptive models are closely associated with the classification of epidemics and epidemic processes. Although they inevitably have some explanatory and predictive properties, they are essentially summaries of the processes under study. Two sets of descriptive model based upon a dichotomous classification of one parameter have been widely used in botanical epidemiology: classification of epidemics into two types according to whether or not the causal organism has more than one cycle of infection within a season (Van der Plank, 1963); and classification of infection by soil-borne parasites into 'rhizoplane' and 'rhizosphere' infection according to the maximum distance from which infection can occur (Baker, 1971). Each of these classifications has excited controversy concerning the underlying mathematical and biological assumptions of the models (Pfender, 1982; Gilligan, 1983). Natural systems rarely submit to division into mutually exclusive categories and so interest is shifting to models that permit more continuous classification (Rouse, Chapter 2; Gilligan, Chapter 4). Multivariate classification, using statistical techniques, is an alternative descriptive strategy (Kranz, 1968) that takes account of more of the natural variation exhibited amongst epidemics. It serves as a preliminary tool in the study of the comparative anatomy of epidemics (Kranz, 1978).

There has long been an interest in prediction of disease and concomitant crop loss (see the reviews of Miller and O'Brien, 1957; Waggoner, 1960). Many of these early models were empirical; a notable example being the Beaumont period for forecasting epidemics of potato blight (Beaumont, 1947). Latterly, interest has switched to mathematical models (Fry and Fohner, Chapter 7; Teng, Chapter 8). This change of emphasis reflects improvements in computing software (Butt and Royle, 1974; Mogk, 1974), and in methods for automatic recording of environmental variables (Pennypacker, 1978). The technology for measurement of biotic variables, especially disease severity, still lags far behind that for abiotic variables, but here too there are promising new developments (Lindow, 1983).

Two significant trends emerge from recent work on the construction of mathematical models for prediction. First is the thesis (Butt and Jeger, Chapter 9) that models should be designed and implemented *ab initio* to suit the needs of the grower. Thus economic (Carlson, 1970; Carlson and Main, 1976) as well as biological and physical constraints must be considered in the design of predictive systems. A second trend is apparent in the re-appraisal of the statistical bases of experimental design for the development and testing of models (Teng, Chapter 8). Traditionally, plant pathologists have borrowed experimental designs from agronomists for field experimentation. The tenet of randomisation, in which treatments are assigned to plots at random, reigns supreme in these designs (Fisher, 1970). Problems arise, however, due to patchiness of inoculum (Lin *et al.*, 1979), interference between adjacent plots with wind-dispersed inoculum (Van der Plank, 1963) and imprecise and biased methods of sampling (James and Teng, 1979). Such difficulties suggest that a re-examination is merited of alternative methods of experimental designs such as systematic designs (Dyke and Shelley, 1976; Jenkyn *et al.*, 1979) and of covariance analysis (Bartlett, 1978), together with a reappraisal of plot design (James and Shih, 1973) and sampling methodology (James and Teng, 1979; Lin *et al.*, 1979; Rouse *et al.*, 1981).

Neither descriptive nor predictive models necessarily distinguish cause from effect. To do so is the purpose of explanatory models (Zadoks, 1971). Two distinct phases, analysis and synthesis, can be distinguished. Analysis is the resolution of an epidemic or some part of an epidemic into a set of hypothesis concerning the nature and behaviour of the components. Typical components are the latent and infectious periods of air-borne pathogens (Hau *et al.*, Chapter 3) and primary and secondary infections of soil-borne pathogens (Gilligan, Chapter 4). The behaviour of the components may be described by a mathematical model. In effect, therefore, analysis provides a set of quantitative 'rules' by which an epidemic, or some part thereof, is believed to behave. Synthesis involves the re-assembling and simulation of the system under investigation, according to the 'rules' derived from analysis. Synthesis tests how well we understand the behaviour of the system. This form of systems analysis has been widely used in ecology (Patten, 1971). Following the first construction of a simulator of plant disease (Waggoner and Horsfall, 1969) the allied strategies of systems analysis and computer simulation have achieved a central position in botanical epidemiology (Hau *et al.*, Chapter 3; McCartney and Fitt, Chapter 5; Zadoks and Rabbinge, Chapter 10).

The rise of the computer simulator in plant pathology, however, has not gone unchallenged (Vanderplank, 1975, 1982). Simulators may become so complex that the modeller is no longer sure of what is happening within the computer (Madden, 1980). Because, moreover, they comprise complex hypotheses, computer simulators are very difficult to validate (Zadoks, 1978; Teng, 1981) by the widely accepted method of falsifying hypotheses (Popper, 1959). The determining factor in the debate between advocates and opponents of systems analysis and computer simulation is the degree of resolution aspired to in the analysis and synthesis. Vanderplank (1982) argues that the extent to which an epidemic is resolved into its components should be conditional upon minimising experimental error. He concludes that synthesis (and by implication analysis) 'can go no further than to leave epidemics with three variables: the initial inoculum, the progeny/parent ratio and the latent period.' Resolution of an epidemic into just three components is elegantly simple. It is suited to the construction of comparatively simple growth equations. However, synthesis in the sense used by Vanderplank (1982) is essentially an abbreviated summary that is arguably closer to descriptive than explanatory modelling. The condition of minimizing error, moreover, is not absolute but must be judged relative to the objectives of the investigation. Greater variance may be tolerated as a cost of increased resolution of a system into its components, for example by the inclusion of host growth and propagule dispersal in an analysis of the system. The restricted view of synthesis presented by Vanderplank (1982) may have advantage in expediency but the case against systems analysis and computer simulation in plant pathology is not proved (see Hau *et al.*, Chapter 3 and Zadoks and Rabbinge, Chapter 10).

The polarization of views between advocates of growth curve analysis and computer simulation is but one of several polarized discussions amongst modellers. Other dichotomies are evident, either consciously or unconsciously, between advocates of pathogen-oriented and host-orientated models, temporal and spatial models, and deterministic and stochastic models. These dichotomies reflect the intuition of investigators, a third constraint that lies between the definition of objectives and the statement of assumptions in the selection of mathematical models.

The predominance of deterministic models in plant pathology derives from the argument of the averaging effect of large population size (Van der Plank, 1975). Few stochastic models of disease have yet been published (Rouse *et al.*, 1980; Bruhn and Fry, 1982; Sall, 1980) so that the relative merits of deterministic and stochastic models have yet to be tested. There are two instances, however, where the estimation and modelling of the variance as well as the mean may be profitable. These are the study of the initiation of epidemics, when the population density of a parasite is low, and the inclusion of spatial pattern in models for disease progress (Gilligan, Chapter 4). Despite the publication of a deterministic threshold theorem for the development of epidemics of plant disease by Van der Plank (1963), little attention has been paid to this important problem. By contrast, medical epidemiology abounds with work on thresholds for the initiation of epidemics (Bailey, 1975). Consideration of stochasticity is closely allied to consideration of the spatial pattern of disease and of dispersal of disease (McCartney and Fitt, Chapter 5). Plant pathologists have long assumed that disease was uniformly and randomly distributed over a host crop. Although this assumption allows for simplification of models, it does not adequately describe systems where aggregation of lesions and diseased plants occurs (Waggoner and Rich, 1981; Gilligan 1982).

Another fundamental change is apparent in the shift of emphasis from pathogen-oriented models that take little account of the dynamics of host growth to host-oriented models. This trend can be seen throughout the spectrum from the construction of fundamental models for population dynamics (Rouse, Chapter 2; Hau *et al.*, Chapter 3; Gilligan, Chapter 4) to the implementation of models for prediction in systems of disease management (Fry and Fohner, Chapter 7; Teng, Chapter 8; Butt and Jeger, Chapter 9; Zadoks and Rabbinge, Chapter 10). Complementing this change of emphasis is the increasing interest in the effects of enhancing genetic diversity within and between crops on the population genetics of pathogen populations (Mackenzie and Royer, Chapter 6). Work on the interdependence of population dynamics and genetics of both host and pathogen is proceeding rapidly (Vanderplank, 1978; Leonard and Czocher, 1980; Kiyosawa, 1982; Barrett, 1983; Wolfe and Caten, 1985).

Of the potential benefits of mathematical modelling to improving the efficiency of control of crop disease, prediction stands foremost. Two schemes, EPIPRE and BLITECAST, both discussed in succeeding chapters, have had substantial and widescale success. Further advances are imminent (Chapters 7–10). Models for population genetics (Mackenzie and Royer, Chapter 6) and for population dynamics (Rouse, Chapter 26; Hau *et al.*, Chapter 3) assist in the rational deployment of genetic resistance (Kiyosawa, 1982) and serve to enhance chemical control (Skylakakis, 1983). Such fundamental approaches, however, are always in competition with empiricism, to which much success in the control of plant disease can be ascribed (Stakman and Harrar, 1957). This is true not only of epidemiology but of other disciplines in plant pathology such as physiology and applied genetics. To measure the success of mathematical modelling of crop disease only in terms of its realised contribution to crop production is unduly limiting. So narrow an assessment overlooks not only the potential of modelling to make future contributions but also the broader philosophical worth of the discipline.

REFERENCES

Anderson, R. M. (1981). *In* 'Theoretical Ecology' (R. M. May, ed.), pp. 318–355. Blackwell, Oxford.

Anderson, R. M. and May, R. M. (1979). *Nature (London)* **280**, 361–367.

Aylor, D. (1978). *In* 'Plant Disease: An Advanced Treatise,' Vol. 2, pp. 159–180. Academic Press, New York.

Bailey, N. T. J. (1975). 'The Mathematical Theory of Infectious Diseases and its Applications,' 2nd Ed., Griffin, London.

Baker, R. (1971). *Phytopathology* **61**, 1280–1292.

Barrett, J. A. (1983). *Phytopathology* **73**, 510–512.

Bartlett, M. S. (1978). *J. R. Stat. Soc. Sect. B.* **40**(2), 147–174.

Beaumont, A. (1974). *Trans. Br. Mycol. Soc.* **31**, 45–53.

Bernouille, D. (1760). *Mém. Math. Phys. Acad. Sci. Paris*, pp. 1–45.

Bruhn, J. A. and Fry, W. E. (1982). *Phytopathology* **72,** 1306–1312.

Butt, D. J. and Royle, D. J. (1974). *In* 'Epidemics of Plant Diseases: Mathematical Analysis and Modelling' (J. Kranz, ed.), pp. 78–114. Springer-Verlag, Berlin and New York.

Carlson, G. A. (1970). *Am J. Agric. Econ.* **52**, 216–223.
Carlson, G. A. and Main, C. E. (1976). *Annu. Rev. Phytopathol.* **14**, 381–403.
Chester, K. S. (1950). *Plant Dis. Rep. Suppl.* **193,** 190–362.
Cochran, W. G. (1936). *J. R. Stat. Soc. Suppl.* **3**, 49–67.
Colwell, R. N. (1956). *Hilgardia* **26**, 223–286.
Dimond, A. E., Horsfall, J. G., Heuberger, J. W., and Stoddard, E. M. (1941). *Bull. Conn. Agric. Exp. Stn. New Haven* **451**, 635–667.
Dyke, G. V. and Shelley, C. F. (1976). *J. Agric. Sci.* **87**, 303–305.
Fisher, R. A. (1970). 'The Design of Experiments,' (14th Ed.). Oliver & Boyd, Edinburgh.
Gause, G. F. (1934). 'The Struggle for Existence.' Williams & Wilkins, Baltimore.
Gilligan, C. A. (1982). *Trans. Br. Mycol. Soc.* **73**, 193–200.
Gilligan, C. A. (1983). *Annu. Rev. Phytopathol.* **21**, 45–64.
Greenwood, M. (1931). *J. Hyg.* **31**, 336–351.
Gregory, P. H. (1945). *Trans. Br. Mycol. Soc.* **28**, 26–72.
Gregory, P. H. (1948). *Ann. Appl. Biol.* **35**, 412–417.
Gregory, P. H. (1973). 'The Microbiology of the Atmosphere.' Leonard Hill, London.
Hassell, M. P. (1976). 'The Dynamics of Competition and Predation.' Arnold, London.
James, W. C. (1974). *Annu. Rev. Phytopathol.* **12**, 27–48.
James, W. C. and Shih, C. S. (1973). *Exp. Agric.* **9**, 63–71.
James, W. C. and Teng, P. S. (1979). *In* 'Applied Biology' (T. H. Coaker, ed.), Vol. 4, pp. 201–267. Academic Press, New York.
Jeffers, J. N. R. (1978). 'An Introduction to Systems Analysis with Ecological Applications.' Arnold, London.
Jeger, M. J. (1982). *Phytopathology* **72**, 1185–1189.
Jenkyn, J. F., Bainbridge, A., Dyke, G. V., and Todd, A. D. (1979). *Ann. Appl. Biol.* **92**, 11–28.
Jones, F. G. W. and Perry, J. N. (1978). *J. Appl. Ecol.* **15**, 349–371.
Jones, F. G. W., Parrott, D. M. and Ross, G. J. S. (1967). *Ann. Appl. Biol.* **60**, 151–171.
Kiyosawa, S. (1982). *Annu. Rev. Phytopathol.* 20, 93–117.
Kleczkowski, A. (1950). *J. Gen. Microbiol.* **4**, 53–69.
Kranz, J. (1968). *Phytopathol. Z.* **61**, 205–217.
Kranz, J. (ed.) (1974a). 'Epidemics of Plant Disease: Mathematical Analysis and Modelling.' Springer-Verlag, Berlin and New York.
Kranz, J. (1974b). *In* 'Epidemics of Plant Diseases: Mathematical Analysis and Modelling' (J. Kranz, ed.), pp. 1–6. Springer-Verlag, Berlin and New York.
Kranz, J. (1978). *In* 'Plant Disease: An Advanced Treatise' (J. G. Horsfall and E. B. Cowling, eds.), Vol. 2, pp. 33–62. Academic Press, New York.
Kranz, J. and Royle, D. J. (1978). *In* 'Plant Disease Epidemiology' (P. R. Scott and A. Bainbridge, eds.), pp. 111–120. Blackwell, Oxford.
Large, E. C. (1952). *Plant. Pathol.* **1**, 109–117.
Leonard, K. J. and Czocher, R. J. (1980). *Annu. Rev. Phytopathol.* **18**, 237–258.
Lin, C. S., Poushinsky, G. and Mauer, M. (1979). *Can. J. Plant. Sci.* **59**, 121–130.
Lindow, S. E. (1983). *Phytopathology* **73**, 1576–1581.
Lotka, A. J. (1925). 'Elements of Physical Biology.' Williams & Wilkins, Baltimore.
Lotka, A. J. (1932). *J. Wash. Acad. Sci.* **22**, 461–469.
McKendrick, A. G. (1926). *Proc. Edinburgh Math. Soc.* **14**, 98–130.
Madden, L. V. (1980). *Prot. Ecol.* **2**, 159–176.
May, R. M. (ed.) (1978). 'Stability and Complexity in Model Ecosystems.' Princeton Univ. Press, Princeton, New Jersey.
May, R. M. (ed.) (1981). 'Theoretical Ecology, Principles and Applications.' Blackwell, Oxford.
May, R. M. and Anderson, R. M. (1979). *Nature* (*London*) **280**, 455–461.

Miller, P. R. and O'Brien, M. J. (1957). *Annu. Rev. Microbiol.* **11**, 77–110.
Mogk, M. (1974). *In* 'Epidemics of Plant Diseases: Mathematical Analysis and Modelling' (J. Kranz, ed.), pp. 55–77. Springer-Verlag, Berlin and New York
Nicholson, A. J. and Bailey, V. A. (1935). *Proc. Zool. Soc. London* **1935**, 551–598.
Patten, B. C. (ed.) (1971). 'Systems Analysis and Simulation in Ecology,' Vol. 1. Academic Press, New York.
Pearl, R. and Reed, L. J. (1920). *Proc. Natl. Acad. Sci. U.S.A.* **6**, 275–288.
Pennypacker, S. P. (1978). *In* 'Plant Disease: An Advanced Treatise' (J. G. Horsfall and E. B. Cowling, eds.), Vol. 2., pp. 97–119. Academic Press, New York.
Pfender, W. F. (1982). *Phytopathology* **72**, 31–32.
Pielou, E. C. (1977). 'Mathematical Ecology.' Wiley, New York.
Popper, K. R. (1959). 'The Logic of Scientific Discovery.' Hutchinson, London.
Rombakis, S. (1947). *Z. Meteorol.* **1**, 359–363.
Rouse, D. I., Nelson, R. R. and MacKenzie, D. R. (1980). *Phytopathology* **70**, 951–954.
Rouse, D. I., MacKenzie, D. R., Nelson, R. R., and Elliott, V. J. (1981). *Phytopathology* **71**, 1015–1020.
Sall, M. A. (1980). *Z. Pflanzenkr. Pflanzenschutz* **87**, 397–403.
Schrödter, H. (1952a). *Angew. Meteorol.* **1**, 154–158.
Schrödter, H. (1952b). *Nachrichtenbl. Dtsch. Pflanzenschutzdienst (Berlin)* **6**(32), 91–92.
Schrödter, H. (1954). *Nachrichtenbl. Dtsch. Pflanzenschutzdienst (Berlin)* **8**(34), 166–172.
Schrödter, H. (1960). *In* 'Plant Pathology: An Advanced Treatise' (J. G. Horsfall and A. E. Dimond, eds.), Vol. 3, pp. 170–223. Academic Press, New York.
Seinhorst, J. W. (1965). *Nematologica* **10**, 87–94.
Seinhorst, J. W. (1970). *Annu. Rev. Phytopathol.* , 131–156.
Skylakakis, G. (1983). *Annu. Rev. Phytopathol.* **21**, 117–135.
Snow, C. P. (1964). 'The Two Cultures and a Second Look." Cambridge Univ. Press, London and New York.
Stakman, E. C. and Harrar, J. G. (1957). 'Principles of Plant Pathology.' Ronald Press, New York.
Teng, P. S. (1981). *Z. Pflanzenkr. Pflanzenschutz* **88**, 49–63.
Thompson, W. R. (1924). *Ann. Fac. Sci. Marseille* **2**, 69–89.
Thompson, W. R. (1930). *Ann. Appl. Biol.* **17**, 641–648.
Van der Plank, J. E. (1946). *Trans. R. Soc. S. Afr.* **31**, 269–278.
Van der Plank, J. E. (1947). *Ann. Appl. Biol.* **34**, 376–387.
Van der Plank, J. E. (1948). *Empire. J. Exp. Agric.* **16**, 134–142.
Van der Plank, J. E. (1949). *Empire. J. Exp. Agric.* **17**, 18–22.
Van der Plank, J. E. (1960). *In* 'Plant Pathology: An Advanced treatise' (J. G. Horsfall and A. E. Dimond, eds.), Vol. 3, pp. 229–289.
Van der Plank, J. E. (1963). 'Plant Diseases: Epidemics and Control.' Academic Press, New York.
Van der Plank, J. E. (1975). 'Principles of Plant Infection.' Academic Press, New York.
Vanderplank, J. E. (1978). 'Genetic and Molecular Basis of Plant Pathogenesis.' Springer-Verlag, Berlin and New York.
Vanderplank, J. E. (1982). 'Host–Pathogen Interactions in Plant Diseases.' Academic Press, New York.
Varley, G. C., Gradwell, G. R., and Hassell, M. P. (1973). 'Insect Population Ecology: An Analytical Approach.' Blackwell, Oxford.
Volterra, V. (1926). *Atti. R. Accad. Naz. Lincei, Mem. Cl. Sci. Fis., Mat. Nat.* **2**, 31–113. [Transl. by Chapman, R. N. (1931). *In* 'Animal Ecology,' pp. 409–448. McGraw-Hill, New York.]
Waggoner, P. E. (1952). *Phytopathology* **42**, 323–328.

Waggoner, P. E. (1960). *In* 'Plant Pathology: An Advanced Treatise' (J. G. Horsfall, and A. E. Dimond, eds.), Vol. 3, pp. 291–313. Academic Press, New York.

Waggoner, P. E. and Horsfall, J. G. (1969). *Bull. Conn. Agric. Exp.. Stn. New Haven* **698**, 1–80.

Waggoner, P. E. and Rich, S. (1981). *Proc. Natl. Acad. Sci. U.S.A.* **78**, 3292–3295.

Wilcoxon, F. and McCallan, S. E. A. (1939). *Contrib. Boyce Thompson Inst.* **10**, 329–338.

Wolfe, M. S. and Caten, C. E. (eds.) (1985). 'Populations of Plant Pathogens: Their Dynamics and Genetics.' Blackwell, Oxford, in press.

Yule, G. U. (1925). *J. R. Stat. Soc.* **88**, 1–58.

Zadoks, J. C. (1971). *Phytopathology* **61**, 600–610.

Zadoks, J. C. (1978). *In* 'Plant Disease: An Advanced Treatise' (J. G. Horsfall and E. B. Cowling, eds.), Vol. 2, pp. 63–96. Academic Press, New York.

2

CONSTRUCTION OF TEMPORAL MODELS: I. DISEASE PROGRESS OF AIR-BORNE PATHOGENS

D. I. Rouse

Department of Plant Pathology, University of Wisconsin—Madison, Madison, Wisconsin, USA

I. INTRODUCTION

A. Functional Growth Models

There are two major approaches that have been taken to modelling plant disease progression. The first approach utilises functional growth models and is exemplified by Van der Plank's work in which the logistic equation was formally introduced as a useful model for describing plant disease epidemics (Van der Plank, 1963). The term functional is used here to refer to mathematical growth models that relate population size as the dependent variable to time as the independent variable. Functional models take the form of a single equation or set of equations that can be manipulated and fitted to data sets with various mathematical techniques. This functional approach can be contrasted with the systems approach exemplified by the complex plant disease computer simulators first developed in the late 1960s (Waggoner and Horsfall, 1969; Waggoner *et al.*, 1970). Computer simulators may utilize functional submodels in sub-routines within the simulator but the simulator itself cannot be represented explicitly as a mathematical function in the usual sense. In this chapter the functional approach to modelling disease progression will be examined at length, simulation models being discussed in the next chapter.

There have been many recent review articles and books discussing various functional population growth models developed by population ecologists, population geneticists and mathematicians (Bailey, 1957; Bartlett, 1960; Bartlett and Hiorns, 1973; Haberman, 1977; Hazen, 1970; Hiorns and Cooke, 1981; Hoppensteadt, 1975; Keyfitz, 1968; May, 1974; Pielou, 1977; Smith, 1974; Wangersky, 1978). Many of these review papers include a discussion of epidemic models. Thus, a large number of models exist awaiting a new application or interpretation by plant pathologists. In general, those models that have been rediscovered by plant pathologists such as the logistic, Gompertz, Bertalanffy, Richards, or Weibull are historically old population dynamics models. They represent the level of mathematical knowledge of plant pathologists, not the current level of knowledge of population biologists. In fact, the elucidation of several epidemiological concepts and problems no doubt awaits the development and/or rediscovery of mathematically more sophisticated functional models such as the models found in the literature just cited.

Along with a review, from a new perspective, of the models currently used to quantify plant disease progression, an attempt will be made in this chapter to introduce a few more complex models and newer techniques that should in the future find their way into common plant disease epidemiological use.

B. Rationale for Developing Functional Growth Models

A utility of functional mathematical growth models is their ability to summarize, in the form of relatively simple mathematical expressions, the relationship between disease and time, making it possible to compare epidemics on a statistical basis. In other words, functional models help to facilitate the analysis of disease progress data from experiments in which a treatment such as a chemical or a resistant cultivar is being compared, under otherwise identical conditions, with other chemical or cultivar treatments or with a control treatment. This ability to allow comparisons to be made between disease progress data sets representing different treatments is a major justification for creating functional models, but not the only one.

Functional models can also be used to address theoretical epidemiological questions. The early medical epidemic models addressed the questions of whether a threshold was necessary for epidemic development, and if so what the size of the threshold was (Hoppensteadt, 1974). Also addressed was the question of what would be the maximum proportion of the population that became diseased (Kermack and McKendrick, 1927). Similar questions are pertinent to plant disease epidemiology. In addition, functional models can address questions dealing with specific aspects of disease progression. For example, why do disease progress curves take on particular shapes? More specific questions include what is the impact of indeterminant lesion growth on the progress of disease over time, or what is the effect of increasing the latent period on disease progression? These questions may have obvious qualitative answers. The longer the latent period the more slowly an epidemic will progress. However, functional models express biological concepts such as latent period in an explicit quantitative language, the language of mathematics, and provide a means of theoretical analysis of the relationships between components of epidemic progression. The formulation and analysis of the behaviour of functional models containing specific components of epidemic progression can give rise to specific biological hypotheses that can be tested.

If the model has been created from a set of assumptions about the biology of disease progression, then it may be possible under carefully specified conditions to test whether or not a particular disease conforms to a particular biologically significant assumption (hypothesis) by determining whether or not that disease conforms to the functional model. A successful test of a hypothesis using conformation of data to a model as the criterion requires all other assumptions inherent in the model to be accounted for by controlled experimentation. This approach is nothing more than the application of modern scientific methods.

II. STATISTICAL MODELS

One approach to functional modelling is to select a model solely on the basis of examination of disease versus time data already collected, disregarding any *a priori* knowledge or assumptions about the disease–time relationships. These models are referred to as statistical or empirical models. In this case various linear or multiple linear regression equations can be tested for their adequacy as descriptors of the data set and an appropriate model selected. If, when plotted, the data appear curvilinear, any one of a number of transformations may provide a means of straightening the relationship so that the least-squares linear regression procedure can be employed. One category of transformations gives rise to what is commonly referred to as polynomial regression, which is a linear or multiple linear regression procedure with each term containing the independent variable raised to a different power, as shown below:

$$Y = b_0 + b_1 t + b_2 t^2 + b_3 t^3 \tag{1}$$

Polynomial regression has been widely used in many fields of science because almost any set of data can be fitted with some form of polynomial regression model.

There are two related problems with using strictly statistical models. First, it is difficult to gain any direct biological insight from the values of the estimated parameters because the parameters do not have any specific biological meaning applied to them. Parameters define a specific curve within the family of curves describable by a particlar model. For example, b_0, b_1 and b_2 in Eq. (1) are parameters. In Eq. (1), representing a polynomial model, data from two experimental treatments may have been found to differ significantly but the different values of, say, b_2 or b_3 may not give any additional insight as to the biological mechanisms causing the difference. Of course, the biological meaning of the differences between the treatments and, therefore, in some way b_2 and b_3, is inherent in the hypothesis being tested by the experimental treatments. The second problem is that statistical models leave in doubt the reliability of predictions outside the range of the data since there is no reason given *a priori* why additional data points may not change the final model selected.

When data are fitted to a purely statistical model then purely statistical criteria are employed to select the most appropriate model among a possibly large number of models examined. The statistical criteria used include the F-test for significance of the regression equation, the F-test for lack of fit, the size of the correlation coefficient, tests for randomly distributed residuals, and the size of the standard errors of the estimated parameters of the regression equation (Neter and Wasserman, 1974). It is possible that several statistical models may fit a set of data equally well, as judged by statistical standards. In that case the simplest model should be selected. Of course, other knowledge about the population or disease might be useful in selecting the final model, but that implies the assignment of some biological meaning or rationale to the model independent of the data the model is currently being used to describe. The model is no longer purely statistical at that point.

III. MECHANISTIC MODELS

A. Continuous Single Equation Models

In reality, functional models are seldom selected entirely on the basis of the examination of already existing data. As Turner *et al.* (1961) have stated, 'It is a fundamental fact of statistical inference that the information contained in an analysis of experimental data is the sum of the *a priori* information built into the statistical model employed and the *a posteriori* information contained in the data set itself.' There is almost always some *a priori* information available that can be usefully incorporated into model development. Turner *et al.* (1961) identify *a priori* information that is common to many growth processes. For example, growth may be a continuous process or it may have a maximum or asymptote. This information alone may influence the type of model selected. In addition, it may be possible to make specific assumptions leading to a specific mathematical function. This leads to an explicit theoretical or biologically mechanistic model that can be written as a difference or differential equation depending on whether the population is assumed to be increasing discretely or continuously.

All mechanistic functional models have a series of implicit simplifying assumptions following from the biologically significant or important assumptions on which the model is based. If the model is to be interpreted mechanistically, these assumptions must be considered. Williams (1972) developed a series of models describing population growth, beginning with the following list of implicit assumptions:

'1. The environment, with respect to all properties that preceptively affect the organisms, is uniform in space.
2. The environment, with respect to all properties that preceptively affect the organisms, is constant in time.
3. Organisms are distributed uniformly in space.
4. All organisms, with respect to their impact on the environment or each other, are identical at any one time throughout the population.
5. All organisms, with respect to their impact on the environment or each other, are identical regardless of age.
6. All organisms, with respect to their impact on the environment and each other, are identical throughout time.'

Although assumption 5 is redundant with respect to assumptions 4 and 6, Williams wanted to emphasise that the models he initially discussed have no age structure.

Williams (1972) argued that a clear statement of implicit assumptions was important to an understanding of the model and its limitations with respect to biological interpretation. An excellent plant pathological illustration of this point has been given by Pfender (1982) concerning simple and compound interest disease. If all of the above assumptions were met, then the shape of the disease progress curve would, according to Van der Plank (1963), allow a distinction to be made between diseases in which only initially present inoculum could cause disease (simple interest), and diseases in which inoculum produced during the course of the epidemic contributed to new disease (compound interest). Several researchers have collected disease progress data under field conditions and, based on the shape of the disease progress curve, concluded that the disease was compound interest. In fact, if only one of several of the implicit assumptions, 1–6 listed above, were violated, the shape of the disease progress curve might appear to fit the compound interest model when the disease was really simple interest or vice versa. In general, a hypothesis cannot be adequately tested by fitting a functional model to a set of data unless all of the implicit assumptions of the model have been accounted for, leaving only the single biologically important assumption to be tested.

1. Two-Parameter Deterministic Models

The addition of an explicit biologically meaningful assumption to the implicit assumptions listed by Williams allows a specific mathematical formulation of population growth. For example, a simple assumption leading to an explicit biologically meaningful model is that population growth rate is proportional to the current size of the population. In continuous form this model is written as

$$\mathrm{d}Y/\mathrm{d}t = rY \tag{2}$$

where $\mathrm{d}Y/\mathrm{d}t$ represents the instantaneous rate of change in population size, r is the proportionality constant and Y is current population size. The expression $(1/Y)\mathrm{d}Y/\mathrm{d}t$ is often referred to as the relative growth rate since it represents rate of change in population size per individual. In the case of the exponential model the relative growth rate is a constant, r.

Solving the differential Eq. (2) with initial condition $Y = Y_0$ results in the exponential growth model

$$Y = Y_0 \exp(rt) \tag{3}$$

The parameters in this model have a biological interpretation, Y_0 representing initial population size, and r representing relative growth rate. It should be noted that r itself does not contain any biological meaning with respect to specific mechanism(s) of how the population increases.

Williams uses the explicit documentation of implicit assumptions to aid in orderly development of more complex models. By deleting assumption 6 and changing the biologically explicit assumption, the logistic model is derived. The new biological assumption is that the relative growth rate of disease decreases linearly with the increase of disease. This model is usually written as

$$\mathrm{d}Y/\mathrm{d}t = rY(1 - Y), \qquad Y = \frac{1}{1 + \left(\dfrac{1 - Y_0}{Y_0}\right)\exp(-rt)} \tag{4a,b}$$

where Y_0 = initial population size and r = intrinsic growth rate. Since disease data are usually expressed as percent of maximum disease, Y is expressed as a proportion of diseased tissue (disease severity) or proportion of infected plant units (disease incidence).

The logistic model, first published by Verhulst (1845), represents the simplest model incorporating density-dependent regulation of population growth (Pielou, 1977; Roughgarden, 1979). Van der Plank (1963) attributed the mechanism of density dependence of polycyclic plant diseases to the occurrence of multiple infections; as disease progresses the chance of a spore falling on already diseased plant tissue becomes larger and larger. Van der Plank recognized that density-dependent regulation of disease dynamics could be achieved by other mechanisms as well as mutiple infections, but he did not elaborate in mathematical terms on these mechanisms. Some of the models presented later in this chapter will address the question of alternative mechanisms of density-dependent population regulation.

The assumption that relative growth rate decreases linearly can be modified in several ways to produce variations on the logistic theme. Some of these variations have been reviewed in the plant pathology literature (Analytis, 1973; Kranz, 1974b; Madden, 1980) and these models will be briefly presented below. The Gompertz model assumes that the specific rate of disease decreases proportionally to the natural logarithm of disease so that

$$\mathrm{d}Y/\mathrm{d}t = rY(-\ln Y), \quad Y = \exp[-c\exp(-rt)] \tag{5a,b}$$

where c is a constant of integration.

Von Bertalanffy (1957) proposed two models that assume intrinsic rate of disease decreases proportionally to the square root of disease,

$$\mathrm{d}Y/\mathrm{d}t = rY(1 - Y^{1/2}), \qquad Y = \left[1 + \left(\frac{1 - Y_0^{1/2}}{Y_0^{1/2}}\right)\exp(-rt)\right] \tag{6a,b}$$

or the cube root of disease,

$$\mathrm{d}Y/\mathrm{d}t = rY(1 - Y^{1/3}), \qquad Y = \left[1 + \left(\frac{1 - Y_0^{1/3}}{Y_0^{1/3}}\right)\exp(-rt)\right] \tag{7a,b}$$

The models presented in Eqs. (3–7) have been commonly used by plant pathologists, principally because the parameters in each case can be interpreted as having biological meaning; Y_0 representing initial disease and r representing the rate of disease increase. Each model is represented by what is known as a separable first-order differential equation (3–7a). As a consequence, each model has an analytical closed form solution (3–7b). Most significantly, each of these models has a transformation associated with it, allowing ordinary linear regression to be performed when fitting the model to a set of data.

While linear regression may be used for estimation of the parameters in each of these models there are conditions that must be met before statistical inference can be applied to the least-squares regression algorithm. For example, the unexplained or random error associated with the estimate of each of the dependent variable values must be independently identically normally distributed. If the variance associated with non-transformed disease estimates is constant, transformed estimates will have non-constant variance and, therefore, the errors associated with estimation of the dependent variable will not be identically distributed. Zar (1968) discussed the use of non-linear regression to eliminate this problem. Additionally, disease estimates for a given point in time are usually correlated with previously estimated amounts of disease. Starkey (1977) described several alternative approaches to modelling disease progression that eliminate this problem of autocorrelation.

Berger (1981) found the Gompertz model provided a better statistical fit than did the logistic model for 113 sets of disease progress data obtained from the literature. These epidemics tended to increase very rapidly early in the development of disease and then approach saturation at a slower rate than predicted by the logistic model. Why should the rate of disease decrease by the logarithm of disease ($\ln 1 - \ln Y$) rather than decreasing linearly ($1 - Y$)? This question will be addressed in Section III,C of this chapter.

2. Models with Three or More Parameters

The logistic model can be generalised in several ways to make three or four parameter growth models. One means of doing this is to represent the variable Y as absolute amount of disease and introduce the parameter K to represent the maximum amount of disease possible. This leads to the following model

$$\mathrm{d}Y/\mathrm{d}t = rY[1 - (Y/K)], \qquad Y = \frac{K}{1 + \dfrac{1 - Y_0}{Y_0}\exp(-rt)} \qquad \text{(8a,b)}$$

This model has not yet been applied to plant pathological data although it may be useful in the instance when plant growth results in a variable K and disease is expressed in absolute units rather than proportions. Turner *et al.* (1969) have analysed this model for K an exponential function of time, and Brewer (1976) has given a general solution of the equation for time varying K. A method of applying the logistic equation to disease data when significant plant growth occurs during the course of an epidemic is given by Kushalappa and Ludwig (1982).

Richards (1959) proposed the following generalisations of the logistic model

$$\mathrm{d}Y/\mathrm{d}t = \frac{r}{1-m}\,Y(1 - Y^{1-m}), \qquad Y = \left[1 + \left(\frac{1 - Y_0^{1-m}}{Y_0^{1-m}}\right)\exp(-rt)\right]^{1/1-m} \tag{9a,b}$$

This model was developed by Richards as a generalised model incorporating each of the two parameter models [Eqs. (3–7)] as special cases. Madden (1980) has reviewed the inter-relationships of these models. Richards did not assign to the parameter *m*, a specific biological interpretation, arguing instead that this model was empirically useful because it fitted many data sets.

Goel *et al.* (1971) introduced a very similar functional model

$$\mathrm{d}Y/\mathrm{d}t = rY(1 - Y^{\theta}), \qquad Y = \left[1 + \frac{1 - Y_0^{\theta}}{Y_0^{\theta}}\exp(-\theta r t)\right]^{-1/\theta} \tag{10a,b}$$

This version of the generalised logistic model has been used in ecological studies where the parameter (θ) was given the biological interpretation of representing the degree of territoriality of individual members of the population (Gilpin and Ayala, 1973).

A somewhat different three-parameter growth model introduced to plant pathologists by Pennypacker *et al.* (1980) is the Weibull model as follows:

$$\frac{\mathrm{d}Y}{\mathrm{d}t} = \frac{c}{b}\left(\frac{t-a}{b}\right)^{c-1} \exp\{-[(t-a)/b]\}^{c}, \qquad Y = 1 - \exp\{-[(t-a)/b]\}^{c} \tag{11a,b}$$

The parameters in this model do not at first seem to have any *a priori* biological significance. However, it has been shown that *a*, referred to as the location parameter, is equivalent to time of disease onset; *b*, the scale parameter, is inversely related to rate of disease progress; and *c*, the shape parameter, determines the shape of the disease progress curve.

This model was introduced to the plant pathology community on the grounds that it was a flexible growth model. Almost any three-parameter model will be more flexible than a two-parameter model and there are a large number of 'flexible' three-parameter functional models that could be used to describe disease progression. It is not clear why the Weibull model is particularly unique in this regard. Madden (1980) suggested the Weibull was superior to the Richards' function because parameter estimation was more efficient for the Weibull model. With the ready availability of 'user friendly' non-linear regression computer software, most three-parameter models similar in form to models 8–11 can be easily fitted to data sets. Furthermore, model 10 has an advantage over the Weibull since it represents an extension of the commonly used logistic model already familar to plant pathologists and has parameters directly interpretable in biological terms.

One of Van der Plank's contributions was his concept that disease control could be achieved fundamentally in two ways: either by reducing initial disease, Y_0, or by reducing the rate of disease progress, r. This result was derived from the fact that a two-parameter model with terms Y_0 and r was adequate to describe disease progression. Is the existence of a three-parameter model describing disease progression useful for expanding the concept of disease control mechanisms? Is it possible that a third parameter indicates a third fundamental means of plant disease control?

Models represented by eqs. (9–11) can each be converted to a four-parameter model by adding a parameter K similar to Eq. (8) representing carrying capacity or maximum disease. Turner *et al.* (1976) have shown that each of the models 2–6 and 8–10 are members of a family of models with a single generic parent model

$$\frac{\mathrm{d}Y}{\mathrm{d}t} = \frac{r}{k^n} Y^{(1-np)}(k^n - Y^n)^{(1+p)} \tag{12}$$

Clearly, there is an even larger number of four-parameter models than three-parameter models and more five-parameter models than four-parameter models, etc. In each case, by adding a parameter, the flexibility of the model is enhanced, so the model can be made to fit a wider variety of disease progress curve shapes. Increased flexibility is accompanied by the loss of mathematical tractability (i.e. ease of solution), although the abundance of computer-aided numerical analysis methods makes loss of mathematical tractability much less of a problem today than it was in the past.

By representing population growth models as the product of several component functions, various generalised forms of growth models can be recognised. The simplest generalisation is to make the rate parameter a function of time so that

$$\mathrm{d}Y/\mathrm{d}t = r(t)\, f(Y) \tag{13}$$

For each of the models presented thus far $r(t) = r$, a constant. Another possibility would be to represent $r(t)$ as a linear function of time so that $r(t) = b_0 + b_1 t$. In the logistic model, this leads to a quadratic curve of logits versus time. Waggoner (1976) has illustrated how the logistic model could be modified to account for environmental periodicity such as diurnal temperature fluctuations by inserting the following function for $r(t)$:

$$r(t) = (1 + \sin 2\,\pi t)k/2$$

A further generalisation of growth curve functions can be made by representing $f(y)$ in Eq. (13) as the product of two functions $g_1(Y) \cdot g_2(Y)$ where g_1 is an increasing function and g_2 is a decreasing function of Y. As Madden (1980) has indicated, g_1 can be interpreted to represent forces causing an increase in population size and g_2 can be interpreted to represent forces limiting the rate of growth or responsible for density-dependent population regulation. It is possible that r could be a function of disease as well (see Section III,C this chapter). Also, g_1 and g_2 could be dependent on time as well as disease. This leads to the final generalised form for continuous deterministic population growth models

$$dY/dt = r(Y,t)\, g_1(Y,t)\, g_2(Y,t) \tag{14}$$

There are a few problems associated with the use of the models discussed thus far. Firstly, in models with several parameters it is more likely that interpretation of the biological meaning of the parameters will be lost because estimates of the parameter values from regression analysis are correlated. The more parameters in the model the greater the chance for correlation between parameters. If estimates of r and Y_0 from the logistic model are correlated when the model is fitted to a set of data, then the size of r is affected by Y_0. One consequence of this is that there may be a range of values of Y_0 and r that allow close to a 'best fit' of the model to the data. A second problem with many of the models just reviewed is that they are not robust. A model is said to be robust if relatively large changes in the value of the parameters are required to have a significant effect on the behaviour of the model. Rouse *et al.* (1981b) observed that r and Y_0 were significantly correlated in a set of 36 disease progress curves of powdery mildew on wheat obtained from different field plots at the same location. These results can be explained by either of the two problems with logistic-type growth curves just mentioned or because failure to satisfy the implicit assumptions invalidated a biological interpretation of the parameter estimates.

One of the criteria of a good model is that the parameters representing biological components of population growth should be capable of being estimated independently of each other (Shoemaker, 1973). If the correlation between parameters is zero when the model is fitted to a set of disease progress data, then independence of estimation is achieved. Alternatively, the initial disease Y_0 could be estimated from a separate set of observations and then used as a known quantity in the model when estimating r from a set of disease progress data. With the two-parameter logistic model, prior estimation of Y_0 would allow r to be estimated alone from the disease progress data. Estimation of Y_0 requires the ability to measure the initial amount of disease in a field plot. This is usually impractical unless a known amount of initial inoculum is placed in the field plot by the investigator. Even then a small error in estimation of Y_0 if its true value is very small can significantly influence estimation of r.

Although the models presented thus far are quasi-mechanistic in the sense that *a priori* set of assumptions led directly to their derivation they are, in fact, used most frequently in an empirical way, without regard to all of the implicit assumptions. This must be the case since very few of the implicit assumptions mentioned earlier in this chapter are accounted for under the common experimental conditions from which disease progress data are collected. Also, these models contain epidemiological parameters that cannot be measured independently.

B. Models with Discrete Time Components

Analogous discrete models exist for each of the continuous models presented thus far. Discrete models allow population or disease to change only at regular intervals of time. These models are represented by difference equations rather than differential equations. For example, the difference equation analogous to the logistic differential equation in Eq. (4) is

$$Y_{t+1} = Y_t + rY_t(1 - Y_t) \tag{15}$$

A time delay or time lag can be introduced into this model to account for latent period. Assume that diseased plant tissue does not become infectious until after one discrete time step; then only diseased tissue at time $t - 1$ contributes to disease at time $t + 1$, thus

$$Y_{t+1} = Y_t + rY_{t-1}(1 - Y_t) \tag{16}$$

This model is a second-order difference equation since disease at time $t + 1$ depends both on disease present at time t and $t - 1$. In contrast, the model represented by Eq. (15) is a first-order difference equation.

It is now possible to allow population or disease to change continuously with time but to have discrete time lags representing, for example, latent period. Such models are represented by differential difference equations. Van der Plank (1963) used this approach to account for both a discrete latent period (p) and infectious period (i). His model is

$$\mathrm{d}Y_t/\mathrm{d}t = R(Y_{t-p} - Y_{t-i-p})(1 - Y_t) \tag{17}$$

This model is a four-parameter model, but it does not have a straightforward analytical solution. However, Van der Plank (1963) did give the analytical solution for the simpler exponential model with a time lag for latent period. Shiyomi (1970) described a similar model accounting for infectious period, but not latent period. His model is

$$\mathrm{d}Y_t/\mathrm{d}t = r(Y_t - Y_{t-i}),\ Y_t = Y_0 \sum_{n=0}^{\infty} (-1)^n r^n \exp\{r(t - ni)[(t - ni)^n/n!]\} \tag{18}$$

Neither Van der Plank's model nor Shiyomi's model lend themselves to analysis of disease progression data. This is in general true of models with discrete components. The approach often taken with such models is computer-aided numerical analysis which is useful for addressing theoretical epidemiological questions. These models do have parameters with direct biological meaning that can be measured independently of each other, such as the latent period.

C. Models Represented by Systems of Equations

Plant pathologists have tended towards using either the relatively simple functional models presented earlier, such as the logistic, or have utilised the very complex computer simulation approach to quantify disease progression. The range of functional models requiring somewhat advanced mathematical treatment in between these two extremes has been avoided. Just as the mathematics become more difficult as parameters are added to single equations and as discrete time lags are incorporated so also the mathematical tools required to deal with systems of equations will be more advanced. Eventually, in an attempt to make the models as biologically realistic as possible, and substitute explicit mechanistic assumptions for the implicit assumptions, computer simulation and systems analysis finally become necessary. However, before succumbing to systems analysis there are many useful results that can be obtained from models consisting of a set of equations.

An important illustration of the use of systems of equations to describe population dynamics is the model originally developed by Lotka (1924) to account for the impact of age structure on population growth. Historically, Kermack and McKendrick (1927) first formulated what has become known in the medical profession as the general epidemic model by representing the rate of change of each of three components of a population (susceptibles, infectives, removals) with a differential equation. An important contribution of these epidemic models in the medical field has been the quantitative investigation of which parameters govern the spread of contagion. Kermack and McKendrick's model illustrated this contribution and also provided one of the early results concerning threshold population size necessary for epidemic buildup. Their system of equations has been extensively analysed and expanded upon. However, their model has never been applied to plant disease epidemics.

In the following models, consisting of systems of rate equations, the authors were interested in the impact of a specific epidemic component or components on the shape of the disease progress curve. Rather than attempt to develop an all-inclusive computer simulation model, each of the authors began with a series of simplifying assumptions and biologically explanatory assumptions that lead to a system of rate equations. The analysis of the rate equations included finding relationships among the biologically meaningful parameters and/or providing explanations of why certain empirical models have been useful. Ideally, this analysis would include an analytical solution so that disease could be representable explicitly as a function of time alone.

Jeger (1982) introduced a model similar in approach to the deterministic medical epidemic model of Kermack and McKendrick (1927). The model distinguished between infectious tissue, s, and total diseased tissue, Y, as follows:

$$\mathrm{d}Y/\mathrm{d}t = k_1 s, \qquad \mathrm{d}s/\mathrm{d}t = k_2(y - s) \tag{19a,b}$$

where k_1 and k_2 are rate parameters. His model reduces to a second-order differential equation that can be solved explicitly. Jeger compared the relationship between his model and Van der Plank's model stated by Eq. (17) and found that k_1 and k_2 were comparable to R and $1/p$ in that model. This model also provided an explicit explanation for the observation sometimes made that disease increased unexpectedly rapidly after the end of a fungicide programme. This result illustrates the useful role theoretical modelling can make to an understanding of epidemics.

William's (1972) implicit spatial uniformity assumptions (3 and 4) are patently false under the conditions of most epidemiological experiments. Waggoner and Rich (1981) explored the impact on non-random infection on the form of the disease progress curve by expressing the rate of change of diseased hosts (N) to change in lesions (n) separately from the rate of change of diseased hosts with time. Their approach utilised the relationship

$$\frac{\mathrm{d}N}{\mathrm{d}t} = \frac{\mathrm{d}n}{\mathrm{d}t} \cdot \frac{\mathrm{d}N}{\mathrm{d}n} \tag{20}$$

where $\mathrm{d}n/\mathrm{d}t$ represents rate of change in number of lesions with time and $\mathrm{d}N/\mathrm{d}n$ represents the rate of increase in number of infected hosts per change in number of lesions. Waggoner and Rich (1981) accounted for non-random distribution of infections by deriving $\mathrm{d}N/\mathrm{d}n = (1 - x)^{(1+1/k)}$ from the negative binomial distribution where k is the parameter in the negative binomial distribution representing degree of aggregation. This result suggests that the logistic model should not adequately describe many diseases known to have a non-random spatial pattern or non-random dispersion of inoculum. Waggoner and Rich (1981) explain the paradox that the logistic model has been adequate despite this shortcoming. When $\mathrm{d}n/\mathrm{d}t = rKk[(1 - x^{-1/k} - 1]$ and $k = 1$ the model reduces to the logistic equation.

A question raised earlier in this chapter was, why does the Gompertz model seem to fit so many sets of disease progress data? Fleming (1983) illustrated one possible explanation by creating a model similar to the logistic but with the rate parameter r a function of disease. Fleming's approach was to separate the pathogen (spore) population dynamics and disease dynamics by using a series of rate equations. The specific functional relationship between r and Y was $r = a/(1 + apY)$ where a was a function of mean daily number of spores and their rate of successful contact with susceptible host sites, and p represented latent period. By incorporating latent period and sporulation explicitly in the model Fleming's model is more 'realistic' than the logistic model. Since the Gompertz model has a curve shape similar to Fleming's model, it may fortuitously fit many sets of disease progress data.

Another explanation for the success of the Gompertz model can be derived from a model consisting of three rate equations accounting for growth in size of lesions (S) as well as number of lesions (N). The simplest form of this model is

$$\frac{dN}{dt} = k_1 Y \qquad \frac{dS}{dt} = k_2 \qquad \frac{dY}{dt} = N\frac{dS}{dt} + S\frac{dN}{dt} \tag{21a,b,c}$$

where $Y = NS$. A form of the model incorporating terms to account for a density-dependent regulation of population growth is

$$\frac{dN}{dt} = k_1 y(1 - Y), \qquad \frac{dS}{dt} = k_2(1 - S), \qquad \frac{dY}{dt} = N\frac{dS}{dt} + S\frac{dN}{dt} \tag{22a,b,c}$$

Analysis of this model reveals that early in the epidemic disease increases more rapidly than would be expected by the logistic model and disease progress curves can be obtained that closely approximate the curve shape of the Gompertz model (D. I. Rouse, unpublished results).

IV. STOCHASTIC MODELS

The term stochastic is equivalent to the term probabilistic. Any model explicitly incorporating random variability can be referred to as a stochastic model. Usually the term stochastic when used as an adjective for a model also implies a dynamic (time-dependent) model. To make this distinction clear the term stochastic process is often used instead of stochastic model. A stochastic process relates the *probability* of Y to time instead of relating Y to time as in deterministic models. An example is the pure birth process.

Let the number of lesions on individual leaves in a field at time t be represented by the random variable $Y(t)$ and $y(t)$ is a realisation of that random variable. Assume a small time interval Δt exists such that no more than one additional lesion will appear in that interval.

In any given interval Δt there are two possibilities: either no new lesions appear or one new lesion appears. Let the probability that a new lesion appears be represented by $\lambda\Delta t$, where λ can be thought of as a rate constant, then

$$p(y, t + \Delta t) = p(y - 1,t)\lambda\Delta t + p(y,t)(1 - \lambda\Delta t) \quad (23)$$

This equation states that the probability of having y lesions at time $t + \Delta t$ is equal to the probability there were $y - 1$ lesions at time t multiplied by the chance a new lesion would appear, plus the probability there were y lesions at time t multiplied by the chance no new lesion appeared. Rearranging the terms in this equation and assuming the time interval Δt approaches zero leads to

$$\lim_{\Delta t \to 0} \frac{p(y,t + bt) - p(y,t)}{\Delta t} = \frac{\mathrm{d}p(y,t)}{\mathrm{d}t} = p(y - 1,t)\lambda - p(y,t) \quad (24)$$

The solution of this differential equation is

$$p(y, t) = \frac{e^{-\lambda t}(\lambda t)^{y}}{y!} \quad (25)$$

This is the Poisson distribution function and thus the process is referred to as the Poisson process. This stochastic model can be generalised by allowing λ to be a linear function of disease. It can be shown that the distribution of lesions on leaves at time t will be negative binomial given $\lambda = a + by$ for $b > 0$ (Bosewell and Patil, 1970).

Rouse *et al.* (1981c) observed a negative binomial distribution of powdery mildew colonies on wheat leaves in the field. One possible explanation for this observation has just been given. We examined the possibility of a different explanation. A series of experiments was conducted to establish the frequency distributions associated with components of rate-reducing resistance and provided a possible biological explanation for the occurrence of the negative binomial based on a statistical mixture of these different distribution functions (Rouse *et al.*, 1980).

Although stochastic modelling has been generally shunned by plant pathologists, the example given above illustrates the usefulness of this approach to modelling as a means of theoretical explanation of field observations. Also, results from analysis of the frequency distribution of disease can be used in deterministic epidemic models as illustrated by Waggoner and Rich's (1981) model discussed earlier in this chapter. A question may remain, however, of exactly what practical value is the stochastic modelling approach? For example, can this approach be successfully used to predict disease development? The following is an example. From infectivity titration experiments it is known that the number of *Pseudomonas syringae* on an individual leaf determines the probability that the leaf will become infected. Let the function $P(Y|n)$ represent the probability a leaf became diseased at time $t + i$ given that there were n pathogenic bacteria present on the leaf at time t. Let the function $P(n)$ represent the probability the leaf had n bacteria at time t. Then for any given leaf in the field

$$P(Y) = P(Y|n)P(n) \tag{26}$$

where $P(Y)$ represents the probability a leaf will become diseased. By summing (or integrating in the continuous case) for all classes of n the proportion of diseased leaves in the field can be estimated from $P(Y)$.

Recently it was found that the distribution of *P. syringae* on individual leaves did not differ significantly from the lognormal distribution (Hirano *et al.*, 1982). Furthermore, both the mean and variance of leaf populations of *P. syringae* fluctuated with time and location. In this situation the mean number of bacteria per leaf may be a poor predictor of disease since the number of leaves with relatively high populations, and that therefore contribute most significantly to subsequent disease development, can only be estimated from knowledge of both mean and variance (Rouse *et al.*, 1981a). This approach to understanding disease dynamics which utilises information from the frequency distribution of variables in the disease process is a departure from the deterministic modelling approach utilising regression analysis that assumes data conform to normally distributed, constant variance criteria.

V. CONCLUSIONS

There is a wide variety of functional growth models useful for describing disease progression. Historically, the models most often used were mechanistic by derivation but used empirically; that is, without regard to the underlying assumptions of the mechanistic derivation. These models had only two parameters and were widely used because they could be adapted to linear regression. Functions with three or more parameters have been more recently suggested as useful epidemic models. The ready availability of non-linear regression software makes these models accessible to the empiricist. Models represented by systems of equations are also useful but, since they require more advanced mathematical skills, they have not been widely used. This is true also for stochastic models. However, both of these approaches have great promise not only for advancing the theory of plant disease epidemiology but also for application to practical problems.

REFERENCES

Analytis, S. (1973). *Acta Phytomed.* **1**.

Bailey, N. T. J. (1957). 'The Mathematical Theory of Epidemics.' Griffin, London.

Bartlett, M. S. (1960). 'Stochastic Population Models.' Methuen, London.

Bartlett, M. S. and Hiorns, R. W. (1973). 'The Mathematical Theory of the Dynamics of Biological Populations.' Academic Press, New York.

Berger, R. D. (1981). *Phytopathology* **71**, 716–719.

Bosewell, M. T. and Patil, G. P. (1970). *In* 'Random Counts in Models and Structures' (G. P. Patil, ed.), pp. 1–13. Pennsylvania State Univ. Press, University Park.

Brewer, J. W. (1976). 'The Analytical Solution of a Time-Varying Logistic Growth Equation.' *IEEE Trans. Syst. Man. Cybernet.* Institute of Electrical and Electronic Engineers, New York.

Fleming, R. A. (1983). *Phytopathology* **73**, 308–312.

Gilpin, M. E. and Ayala, F. J. (1973). *Proc. Natl. Acad. Sci. U.S.A.* **70**, 3590–3593.

Goel, N., Maitra, S. and Montroll, E. W. (1971). *Rev. Mod. Phys.* **43**, 231–276.

Haberman, R. (1977). 'Mathematical Models in Mechanical Vibrations, Population Dynamics, and Traffic Flow.' Prentice-Hall, Englewood Cliffs, New Jersey.

Hazen, W. E. (ed.) (1970). 'Readings in Population and Community Ecology.' Saunders, Philadelphia.

Hiorns, R. W. and Cooke, D. (eds.) (1981). 'The Mathematical Theory of the Dynamics of Biological Populations II.' Academic Press, New York.

Hirano, S. S., Nordheim, E. V., Arny, D. C. and Upper, C. D. (1982). *Appl. Environ. Microbiol.* **44**, 965–700.

Hoppensteadt, F. (1974). *In* 'Thresholds for Deterministic Epidemics, Lecture Notes in Biomathematics, no. 2.' (P. von den Driessche, ed.). Springer-Verlag, Berlin and New York.

Hoppensteadt, F. (1975). 'Mathematical Theories of Populations: Demographics, Genetics and Epidemics.' Soc. Industrial and Appl. Mathematics, Philadelphia.

Jeger, M. J. (1982). *Phytopathology* **72**, 1185–1189.

Kermack, W. O. and McKendrick, A. G. (1927). *Proc. R. Soc. London Ser. A* **115**, 700–721.

Keyfitz, N. (1968). 'Introduction to the Mathematics of Populations.' Addison-Wesley, Reading, Massachusetts.

Kranz, J. (1974a). *In* 'Epidemics of Plant Diseases. Mathematical Analysis and Modelling' (J. Kranz, ed.), pp. 7–50. Springer-Verlag, Berlin and New York.

Kranz, J. (ed.) (1974b). 'Epidemics of Plant Diseases. Mathematical Analysis and Modelling.' Springer-Verlag, Berlin and New York.

Kushalappa, A. C. and Ludwig, A. (1982). *Phytopathology* **72**, 1373–1377.

Lotka, A. J. (1924). 'Elements of Physical Biology.' Williams & Wilkins, Baltimore.

Madden, L. V. (1980). *Prot. Ecol.* **2**, 159–176.

May, R. M. (1974). 'Stability and Complexity in Model Ecosystems.' 2nd Ed. Princeton Univ. Press, Princeton, New Jersey.

Neter, J. and Wasserman, W. (1974). 'Applied Linear Statistical Models.' Irwin, Homewood, Illinois.

Pennypacker, S. P., Knoble, H. E., Antle, C. E. and Madden, L. V. (1980). *Phytopathology* **70**, 232–235.

Pfender, W. F. (1982). *Phytopathology* **72**, 31–32.

Pielou, E. C. (1977). 'Mathematical Ecology.' Wiley, New York.

Richards, F. J. (1959). *J. Exp. Bot.* **10**, 290–300.

Roughgarden, J. (1979). 'Theory of Population Genetics and Evolutionary Ecology: An Introduction.' MacMillan, New York.

Rouse, D. I., Nelson, R. R. and MacKenzie, D. R. (1980). *Phytopathology* **70**, 951–954.

Rouse, D. I., Hirano, S. S., Lindemann, J., Nordheim, E. V. and Upper, C. D. (1981a). *Phytopathology* **71**, 901 (abstr.).
Rouse, D. I., MacKenzie, D. R. and Nelson, R. R. (1981b). *Phytopathol. Z.* **100**, 143–149.
Rouse, D. I., MacKenzie, D. R., Nelson, R. R. and Elliott, V. J. (1981c). *Phytopathology* **71**, 1015–1020.
Shiyomi, M. (1970). *Noyyo Gijutsu Kenkyusho Hokoku A* **17**, 103–116.
Shoemaker, C. (1973). *Math. Biosci*, **16**, 143–175.
Smith, J. M. (1974). 'Models in Ecology.' Cambridge Univ. Press, London and New York.
Starkey, R. E. (1977). 'Analysis of plant disease epidemics. Ph.D. Thesis, The Pennsylvania State University, University Park.
Turner, M. E., Monroe, R. J. and Lucas, H. L. (1961). *Biometrics* **17**, 120–143.
Turner, M. E., Blumenstein, B. A. and Sebugh, J. E. (1969). *Biometrics* **25**, 577–580.
Turner, M. E., Jr., Bradley, E. L., Jr., Kirk, K. A. and Pruitt, K. M. (1976). *Math. Biosci.* **29**, 367–373.
Van der Plank, J. E. (1963). 'Plant Diseases: Epidemics and Control.' Academic Press, New York.
Verhulst, P. F. (1845). *Mem. Acad. R. Belg. (no. 1)* **18**, 1–38.
Von Bertalanffy, L. (1957). *Q. Rev. Biol.* **32**, 217–231.
Waggoner, P. E. (1976). *Ann. N.Y. Acad. Sci.* **287**, 191–206.
Waggoner, P. E. and Horsfall, J. G. (1969). *Bull. Conn. Agric. Exp. Stn. New Haven* **698**.
Waggoner, P. E. and Rich, S. (1981). *Proc. Natl. Acad. Sci. U.S.A.* **78**, 3292–3295.
Waggoner, P. E., Horsfall, J. G. and Lukens, R. J. (1970). *Bull. Conn. Agric. Exp. Stn. New Haven* **729**.
Wangersky, P. J. (1978). *Annu. Rev. Ecol. Syst.* **9**, 189–218.
Williams, F. M. (1972). *In* 'Growth by Instussusception: Ecological Essays in Honor of G. Evelyn Hutchinson' (E. S. Deevey, ed.). Archon Books, Hamden, Connecticut.
Zar, J. H. (1968). *BioScience* **18**, 1118–1120.

3

CONSTRUCTION OF TEMPORAL MODELS: II. SIMULATION OF AERIAL EPIDEMICS

B. Hau, S. P. Eisensmith and J. Kranz

Tropeninstitut, Abteilung Phytopathologie,
Justus-Liebig-Universität
Giessen, Federal Republic of Germany

I. INTRODUCTION

This chapter deals with modelling for the construction of temporal simulators of plant disease epidemics on aerial crop parts. For this purpose an epidemic can be studied as a system with a number of interconnecting elements and components. The epidemic may thus be split up into a number of sub-systems such as the components of the disease cycle or infection chain, or it may itself become a sub-system of a larger system such as an entire agroecosystem.

The bounding of a system should be based on the research objectives. A boundary should be selected so that the environment affects the system, but the system does not affect the environment. From observations of some specific components at one level we can try to *synthesise* models which simulate behaviour at a *higher* level, and add more components if necessary. Alternatively we can *analyse* behaviour at one level and try to deduce how components interact at a *lower* level.

Much has been learned since the first mathematical analysis of plant disease epidemics by Van der Plank (1963) and the introduction of computer simulations by Waggoner and Horsfall (1969). We will try to evaluate briefly the merits of the major schools of thought, proceed to simulation within the context of the systems approach, and mention aspects of the data base needed for the development of simulators. Then we shall deal with the major topic of this chapter, the choice of mathematical functions in the construction of component models. We intend to show how to develop appropriate mathematical expressions for biological components and how to link these up into comprehensive models. The discussion is based mainly on our own modelling experience and will be confined to polycyclic diseases like mildews, rusts and leaf spots. There will be no universal approach, such as one function valid for every case, but only a compendium of possible choices. By this we hope to assist in the generation of new and even more appropriate functions and procedures.

II. SCHOOLS OF THOUGHT

Vanderplank, (1982, pp. 175, 176) commented on the two major schools of thought and approaches in epidemiology: the analysts and the synthesists. In view of the large number of variables that are encountered in the study of epidemics, he feels that 'some degree of synthesis is inevitable,' but ponders just how much is enough. He warns of the 'real danger that the torrent of unfruitful narrative about analysis, unless it abates, may turn the general plant pathologist against epidemiology.' This is certainly a great risk, but is it true that 'the analytic approach has contributed nothing' (Vanderplank, 1982)? We find it hard to agree, although we share Vanderplank's misgivings about another danger, that of experimental errors.

Obviously experimental errors are commonly committed. However, provided that the same standard of professional skill and conscience is applied, the error should be in the same range irrespective of whether and to what degree an epidemiological structure or process is synthesised. With increasing number of sub-models thorough biological understanding and methodology must be brought to bear. The question is not whether to synthesise, and to what extent, or to analyse, but to decide which approach is better suited to the objectives of the study. Synthetic models, such as differential equations, are obviously most useful when the reaction of the epidemic, as expressed in its disease progress curve to a certain stimulus, is to be studied. Possible stimuli are differences in host plant resistance, density of inoculum, fertilizer dosages, spacing, or weather. Although they are field oriented, synthetic models are black-box models. They provide, for example, no information on the nature of host, pathogen and environment interactions (Kranz, 1978). Yet such details may be needed when deciding how a disease can best be controlled. This is especially so when more than one disease or other production constraint have to be controlled, as in integrated pest management. The interactions between components assume great importance in integrated pest management and hence analytical models may be more appropriate.

In contrast to this more field-oriented approach, researchers with ample greenhouse and growth chamber space may prefer to begin modelling components in order to obtain the building blocks for more comprehensive simulation models.

A. Growth Curve School

This school believes that some aspects of population behaviour can be described by a certain simple mathematical equation. In some of these mathematical equations biological meanings can be assigned to the parameters. These types of equations are called 'biological equations' (Madden, 1980). By finding the proper parameter values one can describe, for instance, population growth phenomena. One can then study aspects of growth (rate or maximum area achieved) by comparing respective parameter values. Although the search for 'the' growth function continues (Turner *et al.*, 1976), most researchers feel this equation will never be found. The reason is that growth is the product of many biological and environmental factors, with complex interactions being the rule rather than the exception. A simple growth model like $\mathrm{d}Y/\mathrm{d}t = rY$ means that all these factors are incorporated in one rate parameter, r, the intrinsic rate of increase. To increase model flexibility, equations with more parameters have been proposed. However, these equations suffer from high correlation among the parameters and increased difficulty of finding and measuring parameter values. We further distinguish 'statistical' equations, such as $Y_i = B_1 t_i + B_2 t_i^2 + \varepsilon_i$, on the grounds that no biological meaning can be ascribed to the parameters B_1 and B_2.

1. Biological Equations

There are two main reasons to account for the attraction of plant pathologists to the growth curve school involving 'biological equations': (1) the stimulus of Van der Plank's (1963) classic book which induced many to think in terms of quantitative epidemiology, and (2) the availability of easy-to-use computer software packages which can be used to find parameter values quickly for any of the common growth functions. Evaluating the utility of these functions for many sets of data can be achieved quickly by a comparison of the residual error sums of squares or the coefficients of determination.

Van der Plank (1963) introduced two growth functions, the simple and compound interest equations, to aid plant pathologists in dealing with the complexity of changing populations of pathogens in populations of hosts under the influence of man and environment. His models were used to summarise and distinguish between epidemics resulting from mono- and polycyclic processes. By using transformations, Van der Plank (1963) linearised disease progress curves and calculated logarithmic and apparent infection rates. This work led to the widespread use of '*r*' values as estimates of the infection rates when epidemics were analysed for purposes of comparisons. Infection rates, however, only describe the change between two states of disease intensities over time and the use of these rates leads to 'black-box' interpretations, as does the speedometer in a car. Van der Plank (1975) actually calls *r* 'the speedometer of epidemics.'

There has been obvious abuse of this infection rate, partly because a test of linearity of transformed curves usually is not done. Another source of inapt application has been pointed out by Vanderplank (1982). He states '... an average infection rate estimated grossly over the full period has little use because it does not reflect seasonal and ontogenetic change.' Hence, *r* values should only be calculated for appropriate periods as indicated by the non-transformed disease progress curve.

On the other hand, it became apparent that the equations proposed by Van der Plank (1963) for mono- and polycyclic diseases were not universally suited for all data. Analytis (1973) and others investigated other possible growth functions (Table I). Even though Kranz (1974b), Hau and Kranz (1977), Analytis (1979), Berger (1981), and Luke and Berger (1982) suggested that no growth equation is suited for all the known epidemic patterns, as diagrammed by Kranz (1974a), other workers (Jowett *et al.*, 1974; Pennypacker *et al.*, 1980) continued to search for new equations (Table I).

In spite of their black-box character, growth functions have an obvious advantage in modelling disease progress as the temporal graph of epidemics. Functions of this type are simple, good for descriptive and predictive models, less suited for explanatory models, and easy to develop and to use—as long as their limitations are also understood. They are purely phenological, deterministic, and do not provide clues as to with which phase of the pathogen–host–environment interaction a certain action interferes. For proper disease management such knowledge could be essential, as mentioned above. This limitation also applies to the models based on statistical equations.

Table I. Some epidemiological models derived from growth functions used to describe disease progress curves.

Name	Integrated Equation	Reference	Differential Equation (dY/dt)
Exponential	$a[\exp(rt)]$	Van der Plank (1963)	rY
Mitcherlich (simple interest)	$a[1-\exp(-rt)]$ with $a=1$	Analytis (1973); Van der Plank (1963)	$r(a-Y)$ with $a=1$
Bertalanffy	$a\{1-b[\exp(-rt)]\}$	Analytis (1973)	$r(a-Y)$
Logistic (compound interest)	$a/\{1+b[\exp(-rt)]\}$ with $a=1$	Richards (1959); Van der Plank (1963)	$rY(a-Y)$ with $a=1$
Gompertz	$a(\exp\{-b[\exp(-rt)]\})$	Analytis (1973)	$rY\ln(a/Y)$
Richards	$a\{1-b[\exp(-rt)]\}^d$	Richards (1959)	$drY[(a/Y)^{(1/d)}-1]$
Weibull	$1-\exp\{-[(t-a)/b]^c\}$	Pennypacker *et al.* (1980)	$(c/b)(1-Y)[-\ln(1-Y)]^{(c-1)/c}$

2. *Statistical Equations*

An alternative within this first school to the use of 'biological' growth functions for the description of epidemics is the use of 'statistical' growth functions ranging from linear, polynomial, and multiple linear regression analyses to autocorrelation, first-difference, and non-linear regression models (Table II). Madden (1980) summarized these types of models and pointed to stochastic differential models (Sandland and McGilchrist, 1979) as the way to proceed. Kranz (1974b) listed examples of workers who used non-stochastic models for describing disease progress or estimating infection, incubation, or

Table II. Some statistical models used to describe disease progress in relation to time.

Name	Equation
Linear model	$Y = B_0 + B_1X + e$
Polynomial model	$Y = B_0 + B_1X + B_2X^2 + \cdots + B_nX^n + e$
Multiple linear model	$Y = B_0 + B_1X_1 + B_2X_2 + \cdots + B_nX_n + e$
Second-order quadratic with interaction term	$Y = B_0 + B_1X_1 + B_2X_2 + B_{11}X_1^2 + B_{22}X_2^2 + B_{12}X_1X_2 + e$
Autocorrelation model	$Y = B_0 + B_1Y_{t-1} + B_2Y_{t-2} + \cdots + B_nY_{t-n} + e$
First difference model	$Y_i - pY_{i-1} = a(1-p) + B_1(X_i - X_{i-1}) + e$
Autoregressive model	$Y = B_0 + B_1Y_{t-1} + \cdots + B_nY_{t-n} + D_1X_1 + \cdots + D_mX_m + e$
Non-linear model	$Y = B_0X^{B_1} + e$
Stochastic differential model	$(1/Y)dY/dt = f(a, t) + g(B, t)e_t$

latent periods. But he also noted the need for stochastic models for simulation.

The first realistic models using statistical equations were composed from series of partial regression analyses, for example, to produce the 'negative prognosis' of potato late blight (Schrödter and Ullrich, 1965). Later, when better computer software and hardware became available, full-fledged multiple regression analyses (MRA) were introduced into epidemiology (Kranz, 1968; Royle and Thomas, 1972; Analytis, 1973). Butt and Royle (1974) discussed their applications, merits and disadvantages in some detail.

Stepwise multiple regression analyses have proved to be powerful tools for the analysis of field experiments. They are also helpful in parameter estimations for models, or in determination of functions for sub-routines.

Multivariate linear regression equations have been proposed and stood the test in forecasting. They can also be used to simulate epidemics (Royle and Thomas, 1972; Analytis, 1973), or their components (section IV,B,2). MRA models explain responses in the dependent variables as the sum of linear functions of independent variables, 'with little regard to the manner in which the independent variables exercise their control' (Butt and Royle, 1974). Although the univariate *x*, *y* relationship of the synthetic models is replaced by multiple *x* and *y* interactions, multiple regression models still have limited scope.

Many of the variables included in the analysis are in fact non-linear. MRA simulators are prone to implicit errors, and they may distort important biological facts. Another weakness of MRA simulators is the lack of provisions for unexpected changes in the system, and its inputs. Waggoner (1974) has elucidated this shortcoming of MRA models. For instance, simulators that emanated from MRA could well describe the course of an epidemic according to weather variables without having entered information on the presence or absence of inoculum into the model equation. From our own experience with MRA we know that the factors indispensable for sporulation and infection, e.g. leaf wetness duration, may show as very insignificant variables in stepwise MRA analyses. Hence, simulators based on statistical equations may be unsatisfactory for realistic simulators of the internal events in epidemics which eventually are measured in a disease progress curve or gradient. A '. . . regression model describes at only one level of explanation, that of the end product' (Butt and Royle, 1974). All this justifies more flexible types of analytical models for simulation.

B. Simulation School

The second school of thought is based on a systems approach of synthesis of population behaviour from an understanding of the interacting components that compromise population processes. This methodology, which may lead to complex simulation models containing a collection of component sub-models, has been introduced in various fields related to ecology (Patten, 1971, 1972), including horticulture, agronomy and entomology. The advent of high speed computers and simulation languages has given impetus to this approach in plant pathology.

The first computer simulator in plant pathology that was both analytical and flexible was written for an aerial epidemic (Waggoner and Horsfall, 1969). Such an approach implies both analysis of a number of processes and states, and their synthesis into a comprehensive model. After the publication of this first simulator others have been presented (see Waggoner, 1978). In fact, the disease progress curve as the temporal graph is the only output of most simulators in epidemiology so far. Spatial aspects of an epidemic have been treated in only a few models, e.g., EPIMUL (Kampmeijer and Zadoks, 1977).

Simulators also have shortcomings, and, consequently, critics. Much of the criticism is justified (e.g. Vanderplank, 1975, 1982; Baker, 1982) and modellers should heed them. Some of the shortcomings are due to poor craftsmanship, in combining biology, mathematics and software technology. Others such as error perpetuation and compensation amongst components are inherent in the modelling of complex and continuous dynamic models. To overcome both problems by appropriate testing of sub-models, and their interactions, is a serious challenge to all modellers. Abiding by the rules of systems analysis will help them in their endeavours and in developing their specific skill.

The simulation and growth curve schools are not mutually exclusive. For example, the use of a growth function for lesion growth as a model of a sub-component in a model of a larger system would be a way of combining both schools. We feel that a thorough understanding of the components of the disease cycle cannot be accomplished by using simple summary equations. Growth functions reproduce trends of disease progress in a continuous form and in some cases allow us to extrapolate or interpret 'biological' meaning for the parameters. However, they provide no information on the interactions of various elements from the infection chain or disease triangle. These interactions are what produce the great diversity of disease progress curves and it is these interactions that can be manipulated to manage epidemics as in integrated pest management.

III. THE SYSTEMS APPROACH

Watt (1966) defines a system as 'an interlocking complex of processes characterized by many reciprocal cause–effect pathways.' For our needs we will say that in epidemiology the epidemic, i.e. the interactions of pathogens with hosts under the influence of man and environment, can be regarded as the system of interest.

Within most systems, sub-systems or components exist which are elements of a system as outlined by Kranz (1978). System structure is determined by how these components are interwoven with related variables to form the causal chain between system inputs and outputs. Flow charts, block and relational diagrams are usually employed to help visualize hypothesised system structure.

The systems approach is essentially a problem-solving methodology based on the systems concept (Manetsch and Park, 1974). As a methodology it extends from defining a set of needs to an operational system which satisfies these needs after having been modified by certain resource limitations. Inherent in this approach is the iterative learning process whereby information is accumulated, processed, and tested at many stages which are repeated until the outcome from a particular stage is satisfactory. Models are used in this process as guidelines and yardsticks. No outcome is sacrosanct and may be reevaluated and changed at a later phase.

The main attribute of this approach is that within the disease square (Van der Plank, 1963) states and rates are determined by a network of components, all of them changing over time. The systems approach usually implies the use of quantitative models and often computer simulation. The goal of this approach can range from unravelling interlocking populations dynamics, which leads to an increased understanding of epidemics, to the development of an efficient, reliable system for predicting or managing disease outbreaks. Within the systems approach, systems analysis, systems control and systems design (Ruesink, 1976a) may be distinguished.

In epidemiology, systems analysis is a research strategy or framework for solving complex problems of epidemics, concerning their structural components and their behaviour in a way congenial to biological research. Experiments and models interact at each stage of systems analysis, which leads to a recycling process between both (Kranz and Hau, 1980).

Models are simply abstractions of the real world and modelling is just the process of constructing such abstractions. Kranz and Royle (1978) and Batschelet (1966) have listed some characteristics and types of models. Construction of models forces detailed attention to be given to system structure and inputs. It is sometimes possible to construct analytical (in a mathematical sense) models for which a general solution exists which can be used to relate system outputs to system inputs, e.g. a regression equation.

However, other times simulation models or 'simulators' are employed to compute the time-dependent output for a specific set of model inputs and parameter values. Thus simulation is the process of obtaining specific time solutions of a mathematical model in hopes of mimicking reality, or using a functioning model to test different design or management strategies. Because sometimes models become large and complex, and their intricate details lost in a mass of computer output, a warning is in order: one must never lose sight of the specific assumptions concerning system inputs and system structure upon which the simulator is based. Obviously, 'Model building is as much an art as a science' (Baker and Curry, 1976) which requires creative thinking to translate a complicated real-world agricultural system into a useable simulator.

IV. THE DATA

A. Acquisition of Data

Data are needed to construct models, to test their truthfulness and usefulness and eventually to operate them. Collection of data usually starts in libraries. Beyond the stage of a conceptual model, however, data should be generated specifically to define parameters and estimate parameter values (e.g. constants) for the model under development. Here shopping for information from literature and from often incommensurable experiments should no longer be acceptable. It will soon be discovered that specific biological information is not yet available. Such insufficiencies of the data should be corrected by data from appropriate, preferably holistic, experiments (Kranz, 1972). Whether they are done in growth chambers, phytotrons, greenhouses or in the field is not that relevant, at least not in the early stages of the model building. Eckhardt (1983) demonstrated that EPIGRAM (Aust *et al.*, 1983) simulates a mini-epidemic in growth chambers as well as an epidemic of barley powdery mildew in the field. The difference was in the parameter values and not in the parameters (or components) themselves and hence in the level of simulated disease intensities. The adjustment to parameter values valid for the field may still be done during the development of a model. Adjustment is achieved through verification after each stage of modelling with new sets of experimental data, which thereafter may contribute to further improvement of the model. Construction of simulators appears to be most successful in teamwork with a mathematician (by training or inclination) working together with a group of experimentalists.

The procedure of data acquisition appropriate for consistent modelling has been outlined in detail by Mogk (1974).

Whilst it can be justified to employ tedious and time-consuming experimental techniques for model construction, such an approach is not feasible for model operation. Monitoring of operational inputs has to be straightforward, simple and cheap, as well as substantially accurate. In fact, these monitoring requirements are an essential aspect, at least during the final stages of modelling (Kranz *et al.*, 1984). Practical use and acceptance of a biologically sound and satisfactory simulator will greatly depend on its monitoring requirements. For this reason simulators requiring only weather parameters as inputs have better prospects than those which ask for actual measurements in the field. Host growth stages are equally simple and disease assessment will be superior to spore counts with the present technology. For both parameters adequate sampling techniques are a *conditio sine qua non*.

B. Description of Data

1. Graphical

After data collection has been completed, the next step is to describe the relationships contained within the data. Probably the quickest method is with graphs which show relationships between dependent and independent variables. An easy-to-use interactive computer graphics package can help the researcher to search for relationships. The process can also detect errors that may have occurred during data collection or transformation. Graphs give further clues to the type of statistical methodology that may be applicable. For example, a simple (x,y) plot in which the points occur in a linearly increasing or decreasing fashion may point to the use of linear regression or correlation analyses; an (x,y) plot where the data display an S-shaped pattern may illustrate the need for non-linear regression analyses or data transformation; a three-dimensional (x,y,z) response surface could be described by a multiple linear regression model; a skewed curve might indicate the use of a normal distribution to be inappropriate; and a random scattering may show that no relationship exists. Graphs also highlight areas where the response may be changing quickest, not changing at all, is at a maximum or minimum level, or where more observations are needed.

2. Mathematical

Selection of the appropriate method to use to describe the data mathematically in component models is not easy. Two avenues have been pursued: (1) the construction of a table of (x,y) pairs to find the y value corresponding to a desired x value, or (2) fitting a mathematical function to the data and then using the resulting equation. The first method is quick and easy to implement, especially if a simulator language such as CSMP (IBM, 1975) is used. But errors arise with interpolation and extrapolation. The second method can reduce these errors, but finding the correct function to use and the values of its parameters is certainly neither quick nor easy.

We believe functions used in component models should exhibit properties consistent with biological principles and should allow for extrapolation outside the range of data from which they were constructed, as well as interpolation between data observations. A biologically sensible function must always be preferred to a polynomial even if the latter goes exactly through all the observations. More specifically, we mean that if a leaf expands to a maximum size at which it remains until it falls off the plant, then a non-linear function which has an asymptote is more appropriate than a polynomial function. Another common example is that of using linear regression (section II,A,2) to find an equation for a straight line describing a relationship in the data. Of course, we understand that we should never extrapolate outside the range of the original data, but somehow this is forgotten when the relationship is transformed into FORTRAN statements buried deep inside the computer. The original relationship was for daily mean temperature, but ends up being used with an hourly time step, and hourly temperature data can easily occur outside the original range. The simulation program performs the mathematical evaluation of the linear equation quite correctly but the biological effects of high or low temperatures tend to be decidedly non-linear. It should also be stated that even the proper use of a mean value will not always reflect the true influence of temperature (Schrödter, 1965; Sall, 1980a).

Of all the statistical methodology available to plant pathologists, multiple regression has been the most frequently used technique to construct component models. Non-linear functions have also been used for describing relationships for component models (Analytis, 1980), but have been more widely used to describe disease progress curves (Berger, 1981).

A major problem arises when one attempts actually to fit non-linear functions to data. It is a problem of finding the proper values to use for the parameters of the equation. Excellent software exists to find suitable values; however, starting parameter estimates and partial derivatives are usually required and these may be quite difficult to obtain. Software has been developed by two of us (B.H. and S.P.E.) to alleviate these problems. Statistical methodology also provides stochastic models, but these have yet to be fully utilized by simulation workers (Anderson, 1976).

Component models may consist of 'stochastic generators' which produce variables like time until the next leaf emerges, susceptibility indices of new tissue, virulences of races in a population, or numbers of spores produced from a lesion. Use of stochastic processes may have been avoided earlier because it adds complexity to models, but hopefully gains in model realism will be worth the extra complexity. And of course there will always be room for new ideas for models, e.g. the combination of time and space (Jeger, 1983); or revival of old ideas, e.g. the Law of the Minimum (Waggoner *et al.*, 1980).

V. CONSTRUCTION OF COMPONENT MODELS

We feel that in the past too many words have been devoted to generalities and not enough has been said on the actual use of a specific function for a component model. Therefore, we would like to examine some functions and comment on their utility as mathematical approximations of data. This is an attempt to show how a modeller may proceed in constructing sub-models for the components of a pathogen's life cycle. We will present two examples, the first of which deals with time delays within the infection chain. Examples are latent and incubation periods as well as periods spent by insects or fungi in certain physiological stages. Our second example concerns our ideas on how to handle combined temperature and time effects on physiological processes like germination or lesion growth.

A. Quantification of Development Times

1. Influence of Temperature

The speed of an epidemic is largely influenced by the length of development times (Van der Plank, 1963; Zadoks, 1971). In this section we present some of the functions used to model latent and incubation periods. Because temperature is the most important external factor influencing the duration of development times, much information is available on these phenomena in the literature. However, most of this work was done at constant temperatures. Table III summarises some of the quantitative approaches used to predict development times under the influence of constant temperature.

One can easily see that many different functions have been used to describe latent and incubation periods. For example, the incubation time of barley powdery mildew has been described with a parabola (Aust *et al.*, 1978), a hyperbola (Pauvert and de la Tullaye, 1977; Stephan, 1980), and a five-parameter generalized beta function (Hau, 1983).

The eight functions in Table III can be divided in two main categories. The first contains the hyperbolic and discontinuous linear functions which decrease monotonically and approach an asymptote. All the other functions are U shaped, i.e. they have a minimum point at a certain temperature such that lower or higher temperatures result in longer development times. We believe this last category is more realistic for temperature processes in nature. Nevertheless, authors have used the monotonically decreasing functions with good success for two reasons: First, the entire temperature range was not explored in their work, so those parts of the curve are missing; and second, daily mean temperatures, which rarely exceed the optimum temperature, were used. Much care must be exercised when using such a partial curve in a

Table III. Some functions describing the relationship between temperature (T) and development time DEVT(T).

Function Name[a]		Equation for DEVT(T)	Reference
Hyperbola		a/T^b	Pauvert and de la Tullaye (1977), Stephan (1980)
Discontinuous linear	or	$a/(b + cT) + d$ $(a' + d'T)/(b + cT)$	Zadoks (1971)
Symmetrical catenary	or	$0.5\,a[b^{(T-c)} + b^{-(T-c)}]$ $a \cosh[b'(T - c)]$	Janisch (1925) Analytis (1980)
Second-degree polynomial		$a + bT + cT^2$	Aust *et al.* (1978)
Juxtaposed exponential		$\exp[a + b(1/T)]$ $\quad T \leq T_{opt}$ $\exp[a' - b'(1/T)]$ $\quad T > T_{opt}$	Goloff and Bazzaz (1975)
Third-degree polynomial		$a + bT + cT^2 + dT^3$	Teng (1978)
Asymmetrical catenary	or	$0.5a[b^{(T-c)} + d^{-(T-c)}]$ $a'[\exp(b'T)] + c'[\exp(-d'T)]$	Janisch (1925)
Generalised beta		$a(T - b)^{-c}(d - T)^{-e}$	Hau (1983)

[a] Functions are separated into classes with similar properties. See text for details.

component model, especially when extrapolation would clearly violate biological intuition. We thus dismiss the use of the first two functions and prefer the second group.

Within the second group further division can be made with respect to symmetry (Table III). The second-order polynomial and symmetrical catenary function exhibit symmetry, while the other four functions are asymmetric. The two polynomials will be skipped on the grounds that their parameters have no biological meaning and the extrapolation with third and higher order polynomials is misleading (Hau and Kranz, 1980).

If one examines a series of data sets for the relationship of temperature and development periods as found in Gäumann's (1951) book, it becomes evident that the majority of curves are asymmetric. Shearer and Zadoks (1972) stated that the curve for the latent period of *Septoria nodorum* (Berk.) Berk. on wheat 'has the typical asymmetrical U-shape associated with this type of relationship.' Eversmeyer *et al.* (1980) reported temperature–development curves for the latent period of *Puccinia recondita* Rob. ex Desm. f. sp. *tritici* 'were skewed to the left, with an optimum at 80% of the temperature range.'

Obviously a function which is flexible enough to describe both the symmetric and the asymmetric cases would be more suitable. We have found the beta function useful for describing such relationships. However, finding the best parameter values is non-trivial. The reason for selecting this function instead of the asymmetric catenary function lies with the mathematics of the process at the cardinal temperatures. The mathematical domain of the beta function is restricted to the range (b, d), b and d being the minimum and maximum cardinal points of the process (Heald, 1943). At these temperatures no development occurs, so the development time is indefinite. This idea is supported by data on stripe rust of wheat collected by Tollenaar and Houston (1967). In contrast, the catenary function has definite values at these points, so times can be calculated at or outside of the cardinal points. Keep in mind that these cardinal points are not absolute boundaries due to compensation (Aust *et al.*, 1980). Also, isolates can be found which grow below the minimum, above the maximum, and have other optimum temperatures.

An application of the eight different functions of Table III to the data of Tollenaar and Houston (1967) on the latent period of *Puccinia striiformis* is given in Fig. 1. At 24.5 and 25.5°C only chlorotic flecks after more than 80 days appeared, and, therefore, these data are not used in the approximation

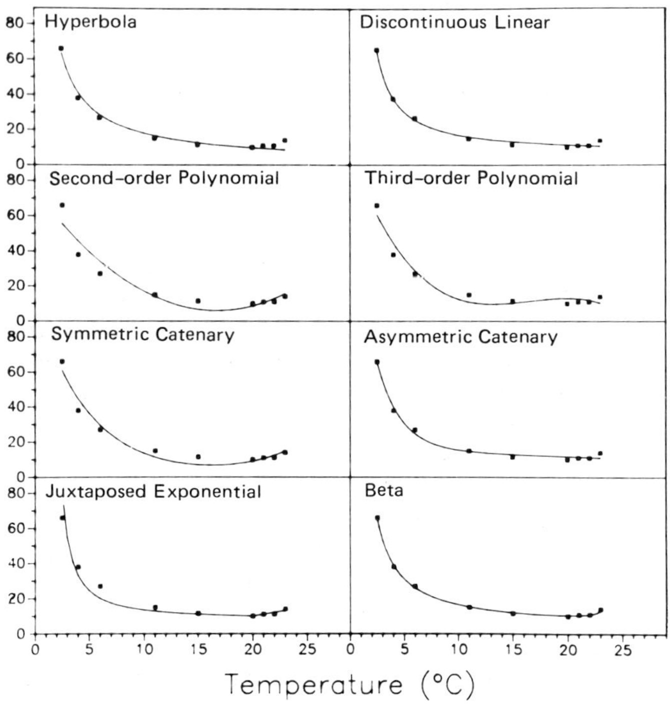

Fig. 1. The eight functions of Table III applied to the data of Tollenaar and Houston (1967) on the latent period of *Puccinia striiformis* in relation to temperature. (Ordinate: latent period in days.)

but they reflect the trend that the latent period increases drastically at high temperatures. It is remarkable that the third-order polynomial would give completely wrong results when extrapolated for temperatures higher than 23°C.

2. Relation to Development Rate

It is fashionable in simulation models to use rates and states according to the following scheme: state at time t plus a small change in time equals the state at time t plus the rate at time t multiplied by the length of the small time change. Be aware that the time step is selected small enough so that the rate can be assumed to be constant during that interval. Instead of examining the time to finish a process, we focus our attention on the speed of development. If we assume the rate is constant, then the development rate is just the reciprocal of the development time. Therefore the development rate DEVR(T) for a certain temperature T, can be calculated as 1.0/DEVT(T), facilitating use of any of the functions of Table III.

In addition to these equations, Table IV contains other proposed rate equations for the effect of temperature on development rate. The first equation is just a logistic growth function applied to each tail of the temperature optimum curve. The second is a more general form of the modified sine function proposed by Schrödter (1965). For his original equation to be used it is first necessary to know the cardinal temperature values and normalise the temperature data into a [0, 1] range using the transformation $T' = (T - T_{min})/(T_{max} - T_{min})$. Also, the rates must be transformed into a [0, 1] interval by dividing the data with the maximum rate. This same type of normalisation was employed by Analytis (1977), and his function can be looked upon as a normalized beta function. The last two functions in Table IV are based on a lethal temperature instead of the maximum where development can still be observed, and do not approach a minimum but exhibit a monotonically decreasing slope towards some base temperature.

Table IV. Some functions to describe the relationship between temperature T and developmental rate DEVR(T).

Equation for DEVR(T)	Reference
$a/\{1 + b[\exp(cT')]\}$ with $T' = \begin{cases} T & \text{for } T \leq T_{opt} \\ 2T_{opt} - T & \text{for } T > T_{opt} \end{cases}$	Stinner *et al.* (1974)
$a \sin^2(b + cT + dT^2 + eT^3)$	Modified after Schrödter (1965)
$a[\exp(bT)] - c[\exp(dT)]$	Modified after Logan *et al.* (1976)
$a/\{1 + b[\exp(cT)]\} - d[\exp(fT)]$	Modified after Logan *et al.* (1976)

Non-linear regression is commonly employed to find suitable parameter values for these equations. Logan *et al.* (1976) explain how starting parameter value estimates can be found for their function, and we have developed software for fitting of the original and modified Schrödter (1965) equations.

3. Procedures with Changing Temperatures

When temperatures remain constant throughout a development period, as in a growth chamber, the development time can be calculated directly using the equations in Table III. In reality, however, temperature never remains constant and we therefore need a way to calculate development times for fluctuating temperatures. Some approaches put forth in the literature will now be addressed (Table V).

The first approach employs the accumulation of heat above a specific base temperature which is the lower limit for development. In this approach a characteristic accumulation of heat units is shown to correspond to a stage of biological development. This facilitates the monitoring of biological development by simply monitoring temperature and accumulating heat units. When the characteristic sum is reached, development is complete. Methods to determine the proper base temperature and to reduce the correctable errors which occur when daily maximum and minimum temperature are used have been published (Arnold, 1959, 1960).

Janisch (1925) criticised this method in the general case because it assumes that a linear relationship exists between development rate and temperature as shown in Fig. 2A. Wang (1960) reviewed the development of the heat unit approach and pointed out more criticisms. Nevertheless, the use of degree–day accumulation has led to some useful applications in predicting insect development (e.g. Foster and Taylor, 1975), and the growth development of various crops (e.g. Ashcroft *et al.*, 1976; Arnold, 1974; Gutierrez *et al.*, 1975; Giese *et al.*, 1975).

Because development rate may not continue to increase with increasing temperature, some modifications of the degree–day calculation formula have been proposed. Baskerville and Emin (1969) introduced horizontal and vertical thresholds for different effects of high temperatures. The first threshold is used in systems where the development rate remains constant above a certain temperature (Fig. 2B). The second threshold allows for no development when the temperature exceeds a certain level (Fig. 2C). It is also possible to allow the effect of temperature above an optimum to decrease until an upper limit is reached (Fig. 2D). Ways to reduce the error of heat accumulation are to employ a non-linear method for describing the temperature/rate of development relationship (Fig. 2E,F), or to measure more frequently, for example, to calculate degree–hours instead of degree–days.

Table V. Some methods to determine actual development time (Z) under fluctuating temperature conditions. DEVT(T) is the function used for constant temperature (see Table III).

Method	Formula[a]	Used by
Heat accumulation (degree–day, –hour)	$\sum_{i=1}^{Z} X_i \geq \text{constant}$	
	where $\begin{cases} X_i = 0 & T_i \leq T_{min} \\ X_i = T_i - T_{min} & T_i > T_{min} \end{cases}$	Shrum (1975)
	where $\begin{cases} X_i = 0 & T_i \leq T_{base} \\ X_i = T_i - T_{base} & T_{base} < T_i \leq T_{hor} \\ X_i = T_{hor} - T_{base} & T_i > T_{hor} \end{cases}$	Baskerville and Emin (1969) (horizontal threshold)
	where $\begin{cases} X_i = 0 & T_i \leq T_{base},\ T_i \geq T_{vert} \\ X_i = T_i - T_{base} & T_{base} < T_i < T_{vert} \end{cases}$	Baskerville and Emin (1969) (vertical threshold)
	where $\begin{cases} X_i = 0 & T_i \leq T_{min},\ T_i \geq T_{max} \\ X_i = T_i - T_{min} & T_{min} < T_i < T_{opt} \\ X_i = (T_{max} - T_i)C & T_{opt} \leq T_i < T_{max} \end{cases}$ with $C = (T_{opt} - T_{min})/(T_{max} - T_{opt})$	Shrum (1975)
Calculation of mean temperatures	$\text{DEVT}\left(\frac{1}{Z}\sum_{i=1}^{Z} T_i\right) \leq Z$	Teng (1978)
Summation of temperature equivalents	$\sum_{i=1}^{Z} X_i \geq 1$	Zadoks (1971); Rapilly and Jolivet (1976)
	where $X_i = \begin{cases} 0 & T_i \leq T_{min},\ T_i \geq T_{max} \\ 1/\text{DEVT}(T_i) & T_{min} < T_i < T_{max} \end{cases}$	
Boxcar function without dispersion		De Wit and Goudriaan (1974)

[a] See text for further details.

Two approaches of Table V, the calculation of mean temperature and the summation of temperature equivalent, make use of a known relationship between development time and temperature, normally an asymmetric U-shaped optimum curve (Zadoks and Schein, 1979). The first of these two approaches is essentially the calculation of a running average in which at each time step the mean temperature of the time already passed is used in

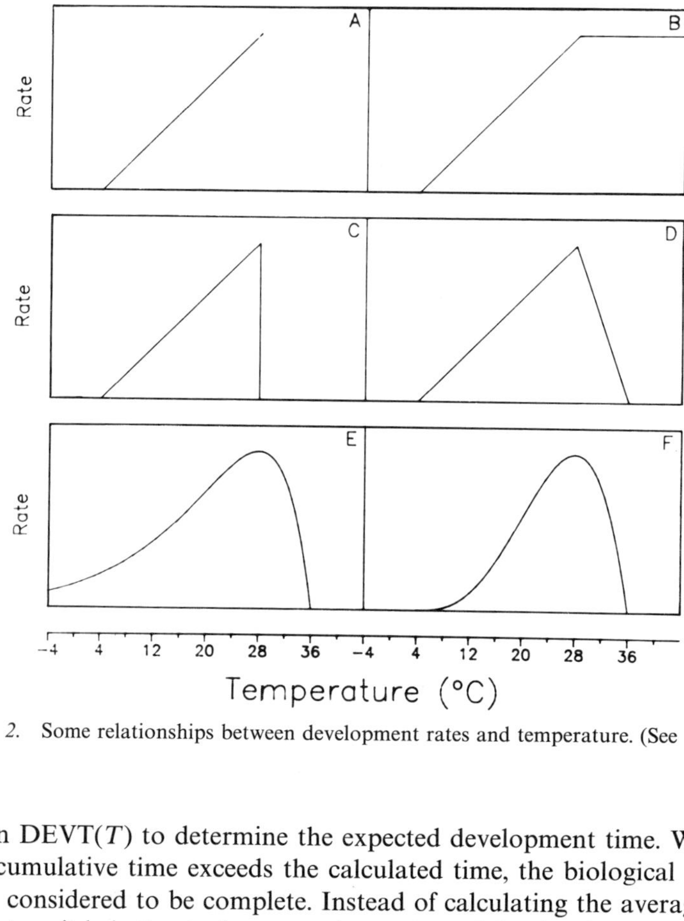

Fig. 2. Some relationships between development rates and temperature. (See text.)

function DEVT(T) to determine the expected development time. When the actual cumulative time exceeds the calculated time, the biological development is considered to be complete. Instead of calculating the average of the temperature it is better to form running means of the development rate and convert it to development times. In this case the U-shaped development time/temperature relationship is inverted to obtain the development rate. This procedure is normally used in such a way that the development rates for each time step, the so-called temperature equivalents, are accumulated until the sum exceeds 1.0. One could think of a third method using running averages, namely the direct calculation of the mean duration. All three methods can lead to different results.

Another approach to determine the end of a process, e.g. incubation, is the use of a technique from computer simulation known as the boxcar train without dispersion (De Wit and Goudriaan, 1974; Waggoner, 1974). In this method the process is divided into certain stages which must be passed during development. All elements starting at the same time will finish the process simultaneously.

In the formulae for the heat accumulation it is assumed that temperatures below the minimum temperature have no effect on the summation. Similarly, for the summation of temperature equivalents the rate of development is set to 0 in this case. High temperatures above the upper threshold are treated in the same way. An alternative would be to assign negative rates to temperatures outside the range in which the biological development takes place. This shows the inhibitory effect of temperatures outside the range on the process. There is some evidence that this occurs in nature. For instance, Waggoner and Parlange (1974a) showed that short periods of lethal temperatures led to a delay of germination and reduced the attainable germination rate. They also reported that the time when the maltreatment occurred had an important influence on the outcome. Similarly, Taylor *et al.* (1955) found that exposure of young sporangia of *Phytophthora infestans* to 40°C from 1 to 20 min induced a lag in the start of direct germination. The two examples support the idea of choosing negative values for low and high temperatures, but it may be difficult to estimate the rates because no development occurs under such constant temperature regimes.

On the other hand, some experimental results in the literature cannot be explained by the procedures of Table V. For example, Waggoner and Parlange (1974b) observed that the spore germination process of *Alternaria solani* spores is faster after chilling. With the same fungus Waggoner and Parlange (1975) showed that the germination time was prolonged when the temperature was frequently changed from 16 to 35°C, though both temperatures had the same development rate. In contrast to this, Goloff and Bazzaz (1975) found that the germination process of the first seed of *Rumex crispus* was accelerated under fluctuating temperature regimes. Therefore, the calculated time using the temperature equivalents exceeded the experimental data drastically.

One major assumption underlying the methods in Table V is that sequence of the temperature values has no influence on the development time. This means that the result will be the same if a series of optimal temperatures are followed by a series of unfavourable temperatures or vice versa. For latent periods of fungal diseases this may not be true because the sub-processes of spore germination, penetration, colonisation, etc., may have different optimum curves (Gäumann, 1951). Therefore, more or less spore germination may occur where spores are wet for 12 hr at 20°C and then for 10 hr at 6°C, than when held at 10 hr at 6°C followed by 12 hr at 20°C, even though the same degree-hour accumulation would result. We will address this point in more detail later.

4. *Several Influencing Variables*

In addition to the effect of temperature, other variables have been reported to influence development rates. Let us consider inoculum concentration as a second variable. Aust *et al.* (1978) proposed a method for the calculation of the incubation period of barley powdery mildew in relation to temperature and spore density. Because inoculum concentration did not change throughout the period, these workers used a different temperature optimum curve for each inoculum concentration. In other words, the temperature/development time relationship is shifted with respect to inoculum density. Let us next consider a second variable like leaf wetness, which is changing throughout the development period. Shearer and Zadoks (1974) used a multiple linear regression equation to relate the latent period of *Septoria nodorum* to daily duration of leaf wetness and daily maximum temperature. Their equation contained terms for the linear and quadratic effects of both variables as well as a linear interaction term. These types of relationships are best viewed as response surfaces in a three-dimensional presentation (Fig. 3).

The last three methods in Table V can be applied to cases with two or more independent variables. There are two approaches. The first is to calculate the mean values of the independent variables and evaluate the equation. Waggoner (1974) has illuminated some of the problems with this approach. The second approach is to use the independent variables in the equation at each time step to calculate a rate of development and then sum these rates until development is complete.

5. *Variation of Development Times*

Many researchers define development periods in terms of time from one stage to another. These definitions may hold for one spore if we define its germination and the resulting lesion produced as a quantal (yes/no) response. While this may be true for a certain individual, even at a constant temperature a population of lesions will not appear at the same time. This is due to inherent variability among individuals and processes. Hence, populations exhibit a mean response with an associated variance creating a sigmoid, cumulative response curve (Zadoks and Schein, 1979).

Because of the inconsistency in the development time between what occurs with an individual and what occurs with a population, we need a way of defining when a development time has been completed for a population. Gäumann (1951) suggested using the point when the rate of colony appearance is maximal. Mogk (1973) proposed other methods to determine the incubation time, such as the mean and the median of the frequency distribution of the appearing colonies or by means of the logit transformation of the cumulative frequency distribution. Shaner (1980) has used probit analysis to determine the latent period for leaf rust on wheat.

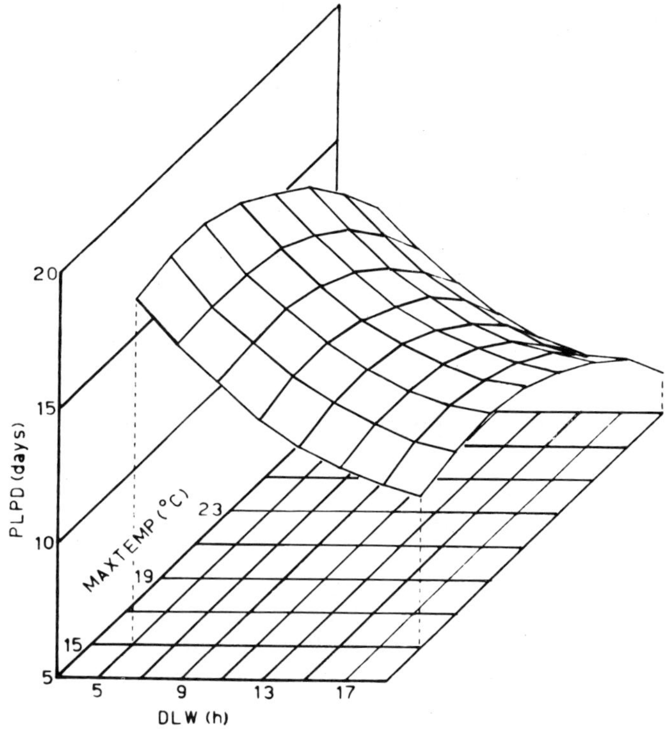

Fig. 3. Response surface showing the effect of duration of leaf wetness (DLW) and maximum temperature (MAXTEMP) on the predicted latent period of *Septoria nodorum* (PLPD). (From Shearer and Zadoks, 1974.)

If we assume a normal distribution can be used to describe the different development times exhibited by the individuals, then we would expect that the distribution function of a normal distribution would approximate spore germination or lesion appearance. Waggoner and Parlange (1974a,b) have presented evidence for just this in the case of *Phytophthora infestans* zoospore germination and Shaner (1980) for the latent period associated with *Puccinia recondita* on wheat. Stinner *et al.* (1975) proposed an empirical function to describe the cumulative germination. This function has properties similar to the normal and the beta distribution functions. Thornley (1977) reported that the gamma distribution is suitable for modelling the distribution of times required for the completion of seed or spore germination, flowering, etc.

Researchers have attempted to incorporate this variation into disease models. One way is to use stochastic distributions to describe the probability associated with each development time. Monte Carlo simulation was used by

Teng (1978) and Sall (1980b) to generate frequency distributions for latent period of barley and of grape powdery mildew, respectively. Another way is to consider the sigmoid response curve of the whole population as a temperature- and time-dependent process which can then be treated as described in the next section.

B. Quantification of Temperature- and Time-dependent Processes

In the preceding pages we have discussed methods which assume that the effect of temperature is independent of the time that it occurs. This means in a steady environment the rate of development is constant during the time course. Now we want to treat cases in which the rate of development of processes varies under the influence of time, even though the temperature remains constant. One example for this kind of process is lesion growth. Another would be the germination of a population of spores.

In many experiments the development in time of a process is observed only at certain intervals and at constant temperatures. The usual way to display these data is either as a two-dimensional figure (Fig. 4) or as a three-dimensional response surface (Fig. 5). A mathematical approximation of the response surface is needed to facilitate estimation of development under temperatures not observed in the experiment. We will discuss different approaches to the mathematical description of the response surface and later

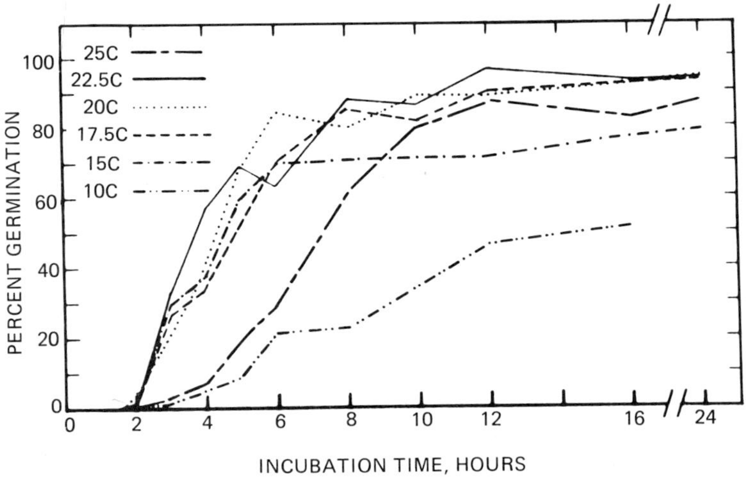

Fig. 4. Percent germination of bean rust urediospores at six temperatures in a two-dimensional graph. (From Imhoff *et al.*, 1981.)

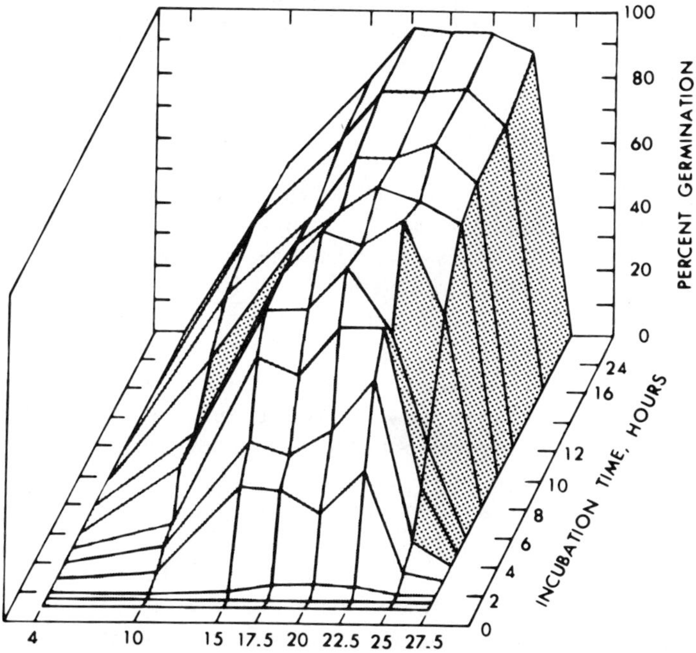

Fig. 5. Percent germination of bean rust urediospores at six temperatures in a three-dimensional graph. (From Imhoff *et al.*, 1981.)

comment on some procedures which can be applied to a variable environment.

1. Mathematical Description of a Response Surface

The procedures to find a mathematical expression for the response surface dependent on time and temperature can be divided into two methods, one involving one stage, the other two. In the one-stage method all parameter values are directly estimated from the original data. In the two-stage method the time component is first analysed separately for each temperature data subset and then calculated parameter values are used in a second analysis where temperature is the independent variable.

Let us start with the two-stage approach, which is easier to handle. Payandeh *et al.* (1978) used this method to develop a model of the germination of *Entomophthora aphidis* resting spores exposed to 12 different temperatures. In the first step for each sub-set of the 12 different temperatures, Richards' (1959) function was applied in the following form:

$$Y(t) = B_1[1 - \exp(-B_2 t)]^{B_3}$$

For the second stage, the estimated values of each of the three parameters were approximated by simple mathematical functions, e.g. polynomials, with temperature as independent variable.

In a similar manner, Imhoff *et al.* (1981) fitted data for bean rust uredospore germination with Richards' (1959) equation, hoping that the resulting parameter values could be expressed as a simple function of temperature. But they could not find a regular pattern for the parameter values, and therefore step functions were employed as approximations. This shows one difficulty which may arise when this method is used. Due to high intercorrelations amongst parameters, especially with Richards' function (Davies and Ku, 1977), the parameter values for one data set can vary over a wide range though the residual sum of squares only changes slightly. It is possible to fix one of the correlated parameters and determine the other one. By means of this procedure it may be easier to find a simple mathematical function to relate parameter values to the independent variable, temperature.

Payandeh *et al.* (1980) pointed out other problems connected with the two-stage regression analysis, e.g. cumulative errors and violation of regression assumptions regarding independent error terms. To overcome these disadvantages they advocate one-stage regression analysis. In an example Payandeh *et al.* (1980) fitted the following function with seven parameters, b_i, in a one-stage procedure using the same data in which T represents temperature:

$$Y(t) = B_1[1 - \exp(-B_2 t)]^{B_3}$$

$$B_1(T) = b_1 T^{b_2} b_3{}^{T b_4}$$

$$B_2(T) = b_5$$

$$B_3(T) = b_6 T^{b_7}$$

To start the non-linear procedures, estimates of starting parameter values are necessary, and it is not trivial to find them. Unfortunately, most authors, e.g. Payandeh *et al.* (1980), give no hints how they obtained starting estimates. One possibility for estimating starting values is to calculate a two-stage regression before the one-step procedure starts.

Another approach to describe the response surface is through the use of multiple linear regression analysis, e.g. a model of the form:

$$Y(t, T) = b_0 + b_1 T + b_2 t + b_3 T^2 + b_4 t^2 + b_5 Tt + \varepsilon$$

Eisensmith and Jones (1981) used this second-order model to develop an equation to detect infection periods of *Coccomyces hiemalis* on sour cherry. One might think of this mathematical description as using a second-degree polynomial for the description of both the temperature and time components. Extensions of this model, i.e. introduction of higher order terms, have been described by Draper and Smith (1966), and computer software contains some sophisticated methodology to find the 'best' combination of these terms. In addition, transformation of both the dependent and independent variable may help improve the fit of this 'statistical' model. Another example of this approach was used by Pearson *et al.* (1977) to model the germination of *Gymnosporangium juniperivirginianae* teliospores in relation to temperature and time.

In cases when the response curve is described by growth functions some of the parameters have biological meaning. In general a process can be characterised by its duration, the maximum level attainable and the course of the development. As a simple example Analytis (1980) proposed the following Mitscherlich (1909) growth function:

$$Y = a[1 - \exp(-kt/X)]$$

for the direct germination of *Phytophthora infestans*. In this equation a and k are constants, t is the time, and X the length of the development time when the process is complete. The duration (X) of the process is a function of temperature and any of the equations of Table III can be used. The asymptote (a) and growth rate (k) of the population are assumed not to vary with temperature. Combining rate k and the duration X to create a new rate $k' = k/X$, means that the effect of temperature is included in the growth rate k' and the duration is omitted from the equation. Looking at the equation from another point of view, we can define a new time variable $t' = t/X$, with t' defined for the inclusive range of 0 to 1. To allow for $y = 1$ when $t' = 1$ it is necessary to set $a = 1/[1 - \exp(-k)]$, i.e. greater than 1. With this kind of time scaling the course of development is identical for each temperature. The only distinction between different temperatures is due to different development times while the asymptote and the course of the process in the transformed system are the same. Therefore, the curves of different temperatures are accommodated by extensions along the time scale. But most time and temperature data we studied do not conform to this hypothesis, because the asymptotic value is not the same. Hence, it is necessary to treat the asymptote as a temperature-dependent function. Hau (1983), for instance, developed a colony growth model for barley powdery mildew from data published by Aust (1981) in which the asymptote and the duration of growth are temperature dependent, but the course of development is temperature invariant in a relative system in which the duration and maximum colony size are normalized to 1.

A third case is possible. In addition to the asymptote and duration, the time course in the relative system may also vary with temperature. A transformation of the equations proposed by Payandeh *et al.* (1978, 1980) in a normalised system would show that the course of the germination in this transformed system is influenced by the temperature. This is due to the fact that in the above equation the parameter B_3 is a function of temperature.

It is possible to describe the data of Payandeh *et al.* (1978, 1980) by Richards' (1959) function in which B_3 is constant and only B_1 and B_2 are temperature functions. As an example we defined B_3 to be 2.0 and assumed that both other parameters are generalised beta functions with temperature as the independent variable. The two cardinal points should be 5 and 27°C. Because of the fact that for B_2 one of the exponents was rather small, we set this parameter to 0 and ended up with the following equations:

$$B_1 = 7.00247(T - 5.0)^{0.490588}\,(27.0 - T)^{0.434918}$$

$$B_2 = 0.01781(T - 5.0)^{0.470795}$$

$$B_3 = 2.0$$

With $R^2 = 0.98$ the fit is slightly better than the equation proposed by Payandeh *et al.* (1980) but the essential difference lies in the fact that in a normalised system the course of the germination is now temperature invariant. To determine the duration of the germination process with the above equations it is necessary to calculate the time when a certain level, for instance, 95% of the maximum, is reached.

2. Procedures in Changing Environment

Using the mathematical description of the response surface one can estimate the response at any time when the temperature remains constant. To account for changing environment different methods have been developed. The simplest method is to calculate a running mean of temperature and then evaluate the surface equation using the corresponding mean temperature at each time step. This is just an extension of the method described for development time. However, this method appears applicable only to situations which have well-defined durations. Otherwise it is not clear when the calculation of mean values is stopped. Jones *et al.* (1980) and Eisensmith and Jones (1981) employed this method in calculating infection periods for apple scab and cherry leaf spot diseases, respectively. In these cases the procedure ended when the leaves dried. Of course, again the assumption is made that the sequence of the temperature values does not matter.

In the second method the mean of the effect of the temperature is calculated instead of the effect of the mean temperature. Therefore, this method is a generalisation of the calculation of temperature equivalents for the determination of development times. If, for instance, the rate of a process is a function of temperature, the running average of the function evaluation at each time is formed and used in the formula for the development in time. Sall (1980a) showed in a simple example that the outcome of both procedures can differ greatly.

Another method for dealing with a fluctuating environment was employed in EPIMAY by Waggoner *et al.* (1972) for modelling germination and sporulation. The increase in germination during the next time period was calculated as follows. Using the temperature value T_c of the coming period, the relevant time function $y(t, T_c)$ for the further development is chosen. If the germination rate already achieved exceeds the maximum value of this function no increase will take place; otherwise, the abscissa of the given germination rate is calculated. Then the function of the same curve is evaluated one time period later to obtain the germination rate after the ongoing time period. Also in this procedure it is not clear when a process is finished. In this procedure the sequence of temperatures influences the outcome, but the increase during a time period is only dependent on the acting temperature and on the level already reached, irrespective of the conditions which led to this level.

A simpler procedure can be applied in cases when the temperature influences only the development time, DEVT, and the asymptote or level, DEVL, of a biological process, but not the course in a normalised system in which both the duration and the level are transformed to a [0, 1] range. The course of development, DEVC, is just a function of the relative duration. The total duration of the process can be determined by summation of the temperature equivalents. Assuming that t is the relative development time already passed, the increase during the next time step with temperature T is therefore

$$\{\mathrm{DEVC}[t + 1/\mathrm{DEVT}(T)] - \mathrm{DEVC}(t)\}\mathrm{DEVL}(T)$$

Again the succession of temperatures is important. The increase is determined by the actual temperature and the relative development time already passed, but the conditions during the development so far have no influence. Hau (1983) applied this method to describe colony growth of *Erysiphe graminis* under fluctuating temperatures.

To show the effect of the four methods described so far we calculated the germination of *Entomophthora aphidis* spores in a changing environment using the function which was developed in the preceding paragraph for constant temperatures. For simplification we assumed that the temperature should be constant at 26°C during the first 30 days and 20°C afterwards. It is obvious from Fig. 6 that the four methods result in quite different germination curves, but the decision as to which behaviour is the correct one can only be made when the changing conditions are tested in experiments.

Another possibility for modelling time- and temperature-dependent processes is the use of boxcar trains. For this it is assumed that a certain number of stages have to be passed through and that the rate of development from one stage to another is constant. De Wit and Goudriaan (1974) describe two

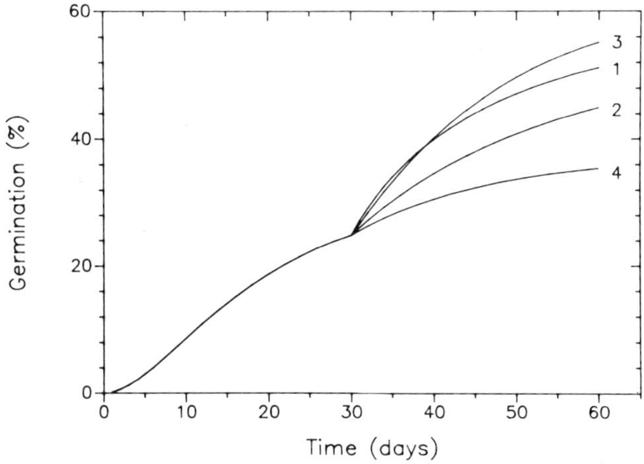

Fig. 6. The germination of *Entomophthora aphidis* spores in changing environment using four different methods: (1) running mean of temperature; (2) running average of the rate and the asymptotic value; (3) method of Waggoner *et al.* (1972); (4) method of Hau (1983). The function applied is based on data of Payandeh *et al.* (1978). (For further explanation, see text.)

methods for the use of boxcar trains to achieve a dispersion of a process. In the first method the number of boxes is fixed for all conditions and therefore the relative dispersion is always constant. The same procedure was adopted by Waggoner and Parlange (1974b) for the germination of *Alternaria solani* spores. It should be mentioned here that the use of the boxcar train with constant relative dispersion can only be applied to cases where the development in the normalised system explained above is temperature invariant. Naturally these processes can also be treated with the previous methods.

In the second boxcar procedure a variable fraction of the contents of each class is shifted, by which a controlled dispersion can be generated. More details on the application of boxcar trains are presented by De Wit and Goudriaan (1974).

The output of the boxcar train has the form of the normal distribution, provided the number of classes is sufficiently high (Goudriaan, 1973). In constrast to this, Thornley (1977) pointed out that the gamma distribution is the correct distribution for a process with several sequential stages. Another kind of boxcar was proposed by Ruesink (1976b), who used a binomial expansion.

3. Several Influencing Variables

In addition to temperature, other variables can affect the time course of a process. In most cases treated in simulators the second factor remains constant during development.

The easiest way to account for the second factor can be applied only to cases where no interactions between the external factors occur and the second factor has only an additive or multiplicative effect on the process. This means, for instance, that the asymptotic value is influenced by the second factor while the shape of the surface, affected by the temperature, is the same. If we regard adult plant resistance as the second factor, the colony growth of barley powdery mildew on the different leaf layers is the same except for the maximum size which can be reached.

If the shape of the surface varies in dependence on the second factor, but this factor can be characterised by a few states only, for instance, light and dark or wet and dry, the response surface can be calculated for each state separately and the procedures explained above can be applied accordingly.

The more general situation in which two influencing factors vary with time is not included in simulators published until now. A possible approach in developing such a function would be a two-step regression procedure in which the parameter values are functions of both external factors.

VI. LINKING OF COMPONENTS

In the preceding pages we have described the construction of components. Now we will deal briefly with the linking of such components. The sequence of the components is given by the underlying biological model, the infection chain of the pathogen. Therefore the transition from one phase to another results from the life cycle in a natural way, for instance by dividing the lesions according to different states, e.g., invisible, sporulating or removed.

Let us consider, as Waggoner (1974) did, the linking between incubation period and lesion growth. When the end of the incubation time is reached, determined by one of the procedures described above, the infections—now visible—are transferred to the next component, the lesion growth. Possibly not all infections beginning the incubation will reach the end, and hence a reduction factor can be introduced to account for the loss during the process. By means of such a factor, which may depend on the environment, the number of new lesions can be corrected.

VII. TESTING OF SIMULATORS

The testing of models is a highly important phase during model development because it is the only way to show the usefulness of a simulator and it is a prerequisite to its application. Testing will be treated here rather shortly because Teng (1981) has discussed its role, methods and problems in an extensive review. Therefore we will confine ourselves to some essential points in connection to the validation of models, i.e. the comparison of model output with the observed reality.

The validity can be tested only with data collected from epidemics in the field which have not been used during model development. Working with component models in which data are normally based on growth chamber experiments, the independent test with field data should cause no problems. The validation either subjective or statistical, should be made with untransformed data. Van der Plank (1975) already criticised misleading figures with logarithmic scales to pretend good agreement.

Kranz (1979) pointed to another problem in connection with the normalisation of model outputs. The reference point at which both the simulated and the observed curve should have the same value can influence the agreement of both curves. The conventional point of reference, the maximum disease intensity, often reflects a too optimistic fit of the simulation. In any case, modellers should be aware that normalisation can be used to compare shapes of curves but cannot detect bias.

A last point which should always be kept in mind is that field observations are affected by random sampling errors which occur independently of errors of bias due to measurement or estimation of disease. Kranz *et al.* (1973) and Kranz (1979) displayed not only the mean disease progress curve of the observed epidemic but also the 90% confidence interval. These intervals may be relatively large and ought to be included routinely with disease progress curves to assist judgements on validation.

VIII. APPLICATION OF COMPONENT MODELS

We have attempted to demonstrate how to model individual components and how to link these up to form simulators. This treatment was determined by systems bounding. Instead of the epidemic itself, components like infection, lesion growth or sporulation may be defined as systems. In such a case any of the individual components may be defined as a simulator in its own right. In fact, an increased usage of individual component models for infection, sporulation, etc., appears to be imminent. They will, for instance, prove to be of great use in forecasting, applied as a tool in more rational crop protection.

Component models which are developed to simulate entire epidemics are a major complement to experimental work in epidemiology (Kranz and Hau, 1980). As conceptual models, computerised or not, their value is mainly heuristic, summarising the present state of knowledge and indicating research needs and approaches. As research goes on, experiments test the model and the model in turn suggests new or different experiments; simulators may develop into guidelines for research projects. This seems particularly valid for long-term undertakings.

Once it has been validated, the simulator may become the 'fourth dimension' in experimentation in addition to field, greenhouse or growth chamber, and laboratory. A reasonably realistic simulator thus as a research tool permits the simulation of all kinds of experiments which otherwise would be impossible, too costly or too hazardous. Moreover, simulators for the epidemics of a given disease may be used for strategic studies, for example to test the effects of changes in agronomic practices in an area on the course of epidemics.

Practically no use has been made of simulators for tactical applications. As 'tactical' we consider the use of simulators for effective short-term forecasts, the evaluation of effects of weather, treatment, cultivars, etc. However, there is certainly scope for such developments as indicated by Schrödter and Aust (1978) for EPIGRAM (Aust *et al.*, 1983).

Simulators of the kind described in this chapter are driven by data monitored in the crop, like weather and spore catches, disease severity, etc. They may be called exogenic simulators. A different type of 'endogenic' simulator was published by Zadoks (1971) for wheat stripe rust and Hau and Kranz (1978) for the effect of the hyperparasite *Darluca filum* on wheat leaf rust. Both models consist of functions as components which summarise existing information with an implicit weather effect. The operator varies parameter values according to assumptions. Changes of disease progress curves are the output from which the sensitivity of the system may be judged.

Some simulators are amenable to use for management games. A number of these have been published recently, e.g. APPLESCAB (Arneson *et al.*, 1978), LATEBLITE (Bruhn *et al.*, 1980) and SPUD CROP (Krawczak and MacKenzie, 1981). However imperfect they may be, they prove to be excellent means to elucidate complex relationships in epidemiology and disease management.

It appears to be no problem to combine simulators for two or more diseases in the same crop, provided all required inputs can be monitored. The high speed of modern computers would make it possible to run simulators consecutively at low cost. But there is an important restriction as diseases compete for infection sites or may affect susceptibility of the host otherwise. Hence, if simulators are run independently they may yield more than 100% disease severity/field. The same would apply to the insect pests that may be present as well. Consequently, such a combination of simulators, as well as other relevant models, is not simply additive. As the relation between diseases, pests, weeds and stress in a crop is more or less interactive, interaction between models must be ensured as well. The simplest way of providing for it would be a common account of infection sites, the proportion that an average lesion of a given disease would occupy, and the rate of competition by each disease or pest for the sites depending on its reaction to environment and host. For weeds this interaction is more indirect through effects on host predisposition, microclimate and collateral host contributions. The joint effects of all these organisms at growth stages of the host sensitive to yield formation can be considered as the total effect of the constraints subsystem on the host. The simulation of all interactions involved requires a higher degree of model integration—and modeller's skill—and a new, more synecological approach in research. It looks confounded but should be as feasible as any in-depth research of an individual pathogen.

It is an intriguing challenge to think of highly integrated component simulators as aids in strategical and/or tactical decision making in integrated pest management. In view of the central importance of the host, plant pathologists must work together with plant physiologists, horticulturalists and agronomists. Because insects destroy the crop and weeds compete for nutrients, water and light, plant pathologists have to realise that their model building efforts must be a part of an inter-disciplinary team.

REFERENCES

Analytis, S. (1973). *Acta Phytomed.* **1**, 1–76.

Analytis, S. (1977). *Phytopathol. Z.* **90**, 64–76.

Analytis, S. (1979). *Phytopathol. Z.* **96**, 156–171.

Analytis, S. (1980). *Z. Pflanzenkr. Pflanzenschutz* **87**, 371–382.

Anderson, J. R. (1976). *Agric. Syst.* **1**, 219–231.

Arneson, P. A., Oren, T. R., Loria, R., Jenkins, J. J., Goodman, E. D. and Cooper, W. E. (1978). 'APPLESCAB: A Pest Management Game.' Michigan State Univ., East Lansing.

Arnold C. Y. (1959). *Proc. Am. Soc. Hortic. Sci.* **74**, 430–445.

Arnold C. Y. (1960). *Proc. Am. Soc. Hortic. Sci.* **76**, 682–692.

Arnold C. Y. (1974). *J. Am. Soc. Hortic. Sci.* **99**, 501–505.

Ashcroft, G. L., Richardson, E. A. and Seeley, S. D. (1976). Appendix D *in* 'Reducing Fruit Losses Caused by Low Spring Temperatures,' A Final report of the Utah Agric. Exp. Stn. to the Four Corners Regional Commission, Proj. No. 562–366-084, Doc. No. 10550101.

Aust, H. J. (1981). *Acta Phytomed.* **7**, 1–76.

Aust, H. J., Hau, B. and Mogk, M. (1978). *Z. Pflanzenkr. Pfalzenschutz* **85**, 581–585.

Aust, H. J., Bashi, E. and Rotem, J. (1980). *In* 'Comparative Epidemiology' (J. Palti and J. Kranz, eds.), pp. 46–56. Pudoc, Wageningen.

Aust, H. J., Hau, B., and Kranz, J. (1983). *Z. Pflanzenkr. Pfanzenschutz* **90**, 244–250.

Baker, C. H. and Curry, R. B. (1976). *Agric. Syst.* **1**, 201–218.

Baker, K. F. (1982). *Annu. Rev. Phytopathol.* **20**, 1–25.

Baskerville, G. L. and Emin, P. (1969). *Ecology* **50**, 515–517.

Batschelet, E. (1966). *BioScience* **16**, 22–24.

Berger, R. D. (1981). *Phytopathology* **71**, 716–719.

Bruhn, J. A., Bruck, R. I., Fry, W. E., Arneson, P. A. and Keokosky, E. V. (1980). 'LATE-BLIGHT: A Plant Disease Management Game.' New York State Coll. of Agric. and Life Sci., Cornell University, Ithaca, New York, Department of Plant Pathology Mimeo 80–1.

Butt, D. J. and Royle, D. J. (1974). *In* 'Epidemics of Plant Diseases: Mathematical Analysis and Modeling' (J. Kranz, ed.), pp. 78–114. Springer-Verlag, Berlin and New York.

Davies, D. L. and Ku, J. Y. (1977). *Biometrics* **33**, 546–547.

De Wit, C. T. and Goudriaan, J. (1974). 'Simulation of Ecological Processes.' Pudoc, Wageningen.

Draper, N. R. and Smith, H. (1966). 'Applied Regression Analysis.' Wiley, New York.

Eckhardt, H. (1983). Biologische Testverfahren für Simulatoren von Epidemien. Dissertation, University of Giessen, Giessen, Fed. Rep. of Germany.

Eisensmith, S. P. and Jones, A. L. (1981). *Phytopathology* **71**, 728–732.

Eversmeyer, M. G., Kramer, C. L. and Browder, L. E. (1980). *Phytopathology* **70**, 938–941.

Foster, J. E. and Taylor, P. L. (1975). *Environ. Entomol.* **4**, 195–202.

Gäumann, E. A. (1951). 'Pflanzliche Infektionslehre.' Birkhäuser, Basel.

Giese, R. L., Peart, R. M. and Huber, R. T. (1975). *Pest Manage. Sci.* **187**, 1045–1052.

Goloff, A. A. and Bazzaz, F. A. (1975). *J. Theor. Biol.* **52**, 259–283.

Goudriaan, J. (1973). *Neth. J. Agric. Sci.* **21**, 269–281.

Gutierrez, A. P., Falcan, L. A., Loew, W., Leipzig, P. A. and van den Bosch, R. (1975). *Environ. Entomol.* **4**, 125–136.

Hau, B. (1983). Epidemiologische Simulatoren als heuristische und operationale Instrumente der Systemanalyse mit besonderer Berücksichtigung eines Modells des Gerstenmehltaus. Dissertation, University of Giessen, Giessen, Fed. Rep. of Germany.

Hau, B. and Kranz, J. (1977). *Phytopathol. Z.* **88**, 53–68.

Hau, B. and Kranz, J. (1978). *Z. Pflanzenkr. Pflanzenschutz* **85**, 131–141.

Hau, B. and Kranz, J. (1980). *In* 'Systems Analysis and Simulation 1980' (A. Sydow, ed.), pp. 405–409. Akademie-Verlag, Berlin.

Heald, F. D. (1943). 'Introduction to Plant Pathology.' McGraw-Hill, New York.

Imhoff, M. W., Main, C. E. and Leonard, K. J. (1981). *Phytopathology* **71**, 577–583.

International Business Machines, Inc. (IBM) (1975). Continuous System Modeling Program III. Program Ref. Man. SH19-7001-2. IBM, Tech. Publ. Dept., White Plains, New York.

Janisch, E. (1925). *Pflügers. Arch. Gesamte Physiol. Menschen Tiere* **209**, 414–436.

Jeger, M. J. (1983). *Plant Pathol.* **82**, 5–11.

Jones, A. L., Lillevik, S. L., Fisher, P. D. and Stebbins, T. L. (1980). *Plant Dis.* **64**, 69–72.

Jowett, D., Browning, J. A. and Haning, B. C. (1974). *In* 'Epidemics of Plant Diseases: Mathematical Analysis and Modeling' (J. Kranz, ed.), pp. 115–136. Springer-Verlag, Berlin and New York.

Kampmeijer, P. and Zadoks, J. C. (1977). 'EPIMUL, a Simulator of Foci and Epidemics in Mixtures of Resistant and Susceptible Plants, Mosaics and Multilines.' Pudoc, Wageningen.

Kranz, J. (1968). *Phytopathol. Z.* **61**, 171–191.

Kranz, J. (1972). *Z. Pflanzenkr. Pflanzenschutz* **79**, 573–581.

Kranz, J. (1974a). *Annu. Rev. Phytopathol.* **12**, 355–374.

Kranz, J. (1974b). *In* 'Epidemics of Plant Diseases: Mathematical Analysis and Modeling' (J. Kranz, ed.), pp. 7–54. Springer-Verlag, Berlin and New York.

Kranz, J. (1978). *In* 'Plant Disease' (J. G. Horsfall and E. B. Cowling, eds.) Vol. 2, pp. 203–222. Academic Press, New York.

Kranz, J. (1979). *Bull. OEPP* **9**, 235–242.

Kranz, J. and Hau, B. (1980). *Annu. Rev. Phytopathol.* **18**, 67–83.

Kranz, J. and Royle, D. J. (1978). *In* 'Plant Disease Epidemiology' (P. R. Scott and A. Bainbridge, eds.), pp. 111–120. Blackwell, Oxford.

Kranz, J., Mogk, M. and Stumpf, A. (1973). *Z. Pflanzenkr. Pflanzenschutz* **80**, 181–187.

Kranz, J., Hau, B. and Aust, H.-J. (1984). *In* 'Pest and Pathogen Control: Strategic, Tactical and Policy Models' (G. E. Conway, ed., pp. 233–241. Int. Ser. Appl. Syst. Anal. Willey-IIASA, Chichester.

Krawczak, D. A. and MacKenzie, D. R. (1981). 'SPUD CROP, The Potato Pest Management Game.' Pennsylvania State Univ., University Park.

Logan, J. A., Wollkind, D. J., Hoyt, S. C., and Tanigoshi, L. K. (1976). *Environ. Entomol.* **5**, 1133–1140.

Luke, H. H. and Berger, R. D. (1982). *Phytopathology* **72**, 400–402.

Madden, L. V. (1980). *Prot. Ecol.* **2**, 159–176.

Manetsch, T. J. and Park, G. L. (1974). Systems analysis and simulation with applications to economic and social systems. Dept. of Electr. Eng. and Syst. Sci., Michigan State Univ. East Lansing.

Mitscherlich, E. A. (1909). *Landwirtsch. Jahrb.* **38**, 537–552.

Mogk, M. (1973). Untersuchungen zur Epidemiologie von *Colletotrichum coffeanum* Noack sensu Hindorf in Kenia, eine Analyse der Wirt-Parasit-Umwelt-Beziehungen. Dissertation, University of Giessen, Giessen, Fed. Rep. of Germany.

Mogk, M. (1974). *In* 'Epidemics of Plant Diseases: Mathematical Analysis and Modeling' (J. Kranz, ed.), pp. 55–77. Springer-Verlag, Berlin and New York.

Patten, B. C. (1971). 'Systems Analysis and Simulation in Ecology,' Vol. 1. Academic Press, New York.

Patten, B. C. (1972). 'Systems Analysis and Simulation in Ecology.' Vol. 2. Academic Press, New York.

Pauvert, P. and de la Tullaye, B. (1977). *Ann. Phytopathol.* **9**, 495–501.

Payandeh, B., MacLeod, D. M. and Wallace, D. R. (1978). *Can. J. Bot.* **56**, 2328–2333.

Payandeh, B., Wallace, D. R. and MacLeod, D. M. (1980). *Can. J. Bot.* **58**, 936–941.

Pearson, R. C., Aldwinckle, H. S. and Seem, R. C. (1977). *Can. J. Bot.* **55**, 2832–2837.

Pennypacker, S. P., Knoble, H. D., Antle, C. E. and Madden, L. V. (1980). *Phytopathology* **70**, 232–235.

Rapilly, F. and Jolivet, E. (1976). *Rev. Stat. Appl.* **24**, 31–60.

Richards, F. J. (1959). *J. Exp. Bot.* **10**, 290–300.

Royle, D. J. and Thomas, G. E. (1972). *Trans. Br. Mycol. Soc.* **58**, 79–89.

Ruesink, W. G. (1976a). *Annu. Rev. Entomol.* **21**, 27–44.

Ruesink, W. G. (1976b). *In* 'Modeling for Pest Management; Concepts, Techniques, and Applications' (R. L. Tummala, D. L. Haynes, and B. A. Croft, eds.), pp. 80–89. Michigan State Univ. Press, East Lansing.

Sall, M. A. (1980a). *Phytopathology* **70**, 338–342.

Sall, M. A. (1980b). *Z. Pflanzenkr. Pflanzenschutz* **87**, 397–403.

Sandland, R. L. and McGilchrist, C. A. (1979). *Biometrics* **35**, 255–271.

Schrödter, H. (1965). *Phytopathol. Z.* **53**, 154–166.

Schrödter, H. and Ullrich, J. (1965). *Phytopathol. Z.* **54**, 87–103.

Schrödter, H. and Aust, H.-J. (1978). *Int. Congr. Plant Pathol. 3rd, Munich, Abstr. Pap.*, p. 310.

Shaner, G. (1980). *Phytopathology* **70**, 1179–1182.

Shearer, B. L. and Zadoks, J. C. (1972). *Neth. J. Plant. Pathol.* **78**, 231–241.

Shearer, B. L. and Zadoks, J. C. (1974). *Neth. J. Plant. Pathol.* **80**, 48–60.

Shrum, R. (1975). *Prog. Rep. Pa. Agric. Exp. Stn.* **347**.

Stephan, S. (1980). *Arch. Phytopathol. Pflanzenschutz* **16**, 173–181.

Stinner, R. E., Gutierrez, A. P. and Butler, G. D., Jr. (1974). *Can. Entomol.* **106**, 519–524.

Stinner, R. E., Butler, G. D., Jr., Bacheler, J. S. and Tuttle, C. (1975). *Can. Entomol.* **107**, 1167–1174.

Taylor, C. F., Smoot, J. J., Quinn, D. O., Rohde, R. A. and Elliott, E. S. (1955). *Phytopathology* **45**, 673–675.

Teng, P. S. (1978). System modelling in plant disease management (Barley leaf rust as a case study. Ph.D. Thesis, University of Canterbury, Canterbury, New Zealand.

Teng, P. S. (1981). *Z. Pflanzenkr. Pflanzenschutz* **88**, 49–63.

Thornley, J. H. M. (1977). *Ann. Bot.* (*London*) **41**, 1363–1365.

Tollenaar, H. and Houston, B. R. (1967). *Can. J. Bot.* **45**, 291–307.

Turner, M. E., Bailey, E. L., Kirk, K. A. and Pruitt, K. M. (1976). *Math. Biosci.* **29**, 367–373.

Van der Plank, J. E. (1963). 'Plant Diseases, Epidemics and Control.' Academic Press, New York.

Van der Plank, J. E. (1975). 'Principles of Plant Infection.' Academic Press, New York.

Vanderplank, J. E. (1982). 'Host–Pathogen Interactions in Plant Disease.' Academic Press, New York.

Waggoner, P. E. (1974). *In* 'Epidemics of Plant Diseases: Mathematical Analysis and Modeling' (J. Kranz, ed.), pp. 137–160. Springer-Verlag, Berlin and New York.

Waggoner, P. E. (1978). *In* 'Plant Disease' (J. G. Horsfall and E. B. Cowling, eds.), Vol. 2, pp. 203–222. Academic Press, New York.

Waggoner, P. E., and Horsfall, J. G. (1969). *Bull. Conn. Agric. Exp. Stn. New Haven* **698**.

Waggoner, P. E. and Parlange, J. Y. (1974a). *Phytopathology* **64**, 605–610.

Waggoner, P. E. and Parlange, J. Y. (1974b). *Phytopathology* **64**, 1192–1196.

Waggoner, P. E. and Parlange, J. Y. (1975). *Phytopathology* **65**, 551–553.

Waggoner, P. E., Horsfall, J. G. and Lukens, R. J. (1972). *Bull. Conn. Agric. Exp. Stn. New Haven* **729**.

Waggoner, P. E., Norvell, W. A. and Royle, D. J. (1980). *Phytopathology* **70**, 59–64.

Wang, J. Y. (1960). *Ecology* **41**, 785–790.

Watt, K. E. F. (1966). *In* 'Systems Analysis in Ecology' (K. E. F. Watt, ed.), pp. 1–14. Academic Press, New York.

Zadoks, J. C. (1971). *Phytopathology* **61**, 600–610.

Zadoks, J. C. and Schein, R. D. (1979). 'Epidemiology and Plant Disease Management.' Oxford Univ. Press, London and New York.

4

CONSTRUCTION OF TEMPORAL MODELS: III. DISEASE PROGRESS OF SOIL-BORNE PATHOGENS

Christopher A. Gilligan

Department of Applied Biology, University of Cambridge, Cambridge, England

I. INTRODUCTION

The progress of disease of soil-borne pathogens might be thought of as a rather slow process with little dynamism to excite the attention of epidemiologists. This view is fostered by two misconceptions. The first is that epidemics must be catastrophic and rapid. The second is that epidemics of soil-borne pathogens are characteristically monocyclic while those of aerial pathogens are polycyclic. Kranz (1974) defined an epidemic as any change in plant disease in a host population in time and space. Epidemics need not, therefore, necessarily be catastrophic; the only requirement is that the amount of plant disease should change, i.e. that it be dynamic. The scale may be different from that of air-borne pathogens, but the progress of disease of soil-borne pathogens is certainly dynamic, as infections occur and spread through root systems. Some soil-borne pathogens are monocyclic such as the cyst nematode, *Globodera rostochiensis*, but arguably, the majority are oligocyclic, achieving several cycles of infection within a single season, and some, e.g. *Phytophthora* spp., are polycyclic.

The infection chain (Gäumann, 1951) is central to the epidemiology and modelling of soil-borne pathogens and is used to structure much of the discussion in this chapter. Certain differences in emphasis exist between the infection chains of aerial and subterranean fungal pathogens (Gilligan, 1985b). Whereas the latent and infectious periods assume almost overriding importance in the 'classical' infection chains of aerial pathogens, they are much less important for soil-borne pathogens. By constrast, inoculum density, increase in the size of lesions or in the extent of infection, limitations on dispersal and the balance between parasitic and saprophytic activities assume great importance in the epidemiology of many soil-borne fungi. Accordingly, the infection chain of many soil-borne fungi can be considered as a heterogeneous chain consisting of several dissimilar phases or cycles (Fig. 1). Distinction is made between primary infection, which is initiated from inoculum in the soil, and secondary infection, which arises from primary infection. Two sorts of secondary infection can, in principle, be distinguished (Gilligan, 1985b): reinfection of the same infection court and cross-infection to another infection court.

Using the infection chain as a link between mathematics and the underlying biology, I intend to examine some of the methods available for modelling the progress of disease of soil-borne pathogens. The choice of strategy for modelling disease progress of soil-borne pathogens is, at the extreme, as equally dichotomous as that for aerial pathogens, considered in the preceding chapters: either the modeller summarises the progress of disease in the parameters of one of two equations, the growth curve approach, or the components of the infection chain are analysed and modelled separately. Computer simulation may then be used to link the models for the infection chain. That is the simulation approach. The mathematical and computing techniques used in either approach may overlap, as, for example, in the use of

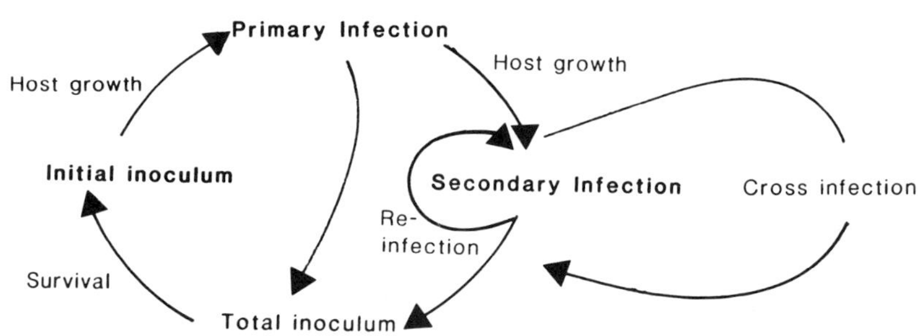

Fig. 1. A general scheme for the infection chain of soil-borne fungi.

numerical integration to solve certain growth equations. The essential distinction between the two strategies is in the degree of detail aspired to in the models. I propose to consider both strategies in this chapter but within each to concentrate upon explanatory models that analyse and, where appropriate, synthesise, rather than merely describe, disease progress. Two themes emerge repeatedly in the discussion. First is that modelling of host growth must occupy a central role in modelling disease progress. Second is that in subterranean systems where movement and dispersal is restricted, consideration must also be given to the spatial pattern of inoculum and infection.

A glossary of the principal random variables and parameters referred to in the text is given in the Appendix.

II. GROWTH CURVE ANALYSIS

A. Standard Non-Linear Models

I begin with standard non-linear models since these are the most commonly used of the growth curve models. The models include the monomolecular, logistic, Gompertz, Weibull, Richards and Bertalanffy functions. The general form of the models is given, after Turner *et al.* (1976), as

$$\mathrm{d}Y/\mathrm{d}t = rf(0,Y)g(Y,K)$$

where $f(0, Y)$ is the generalized distance function between the origin and disease level (Y) that tends to speed epidemics up and $g(Y, K)$ is a generalised distance function between the current level and the maximum level of disease (K) that tends to slow epidemics down; r is a constant of proportionality. The principal standard non-linear models are discussed by Rouse (Chapter 2). The major attraction for plant pathologists in the use of standard non-linear models is that biological meaning, such as initial amounts of disease and rate of increase of disease, can be ascribed to the parameters. It should be noted, however, that much of the attraction of these models lies for population dynamicists in the equilibrium behaviour of the models (see e.g. Rubinow, 1975; Pielou, 1977). The stability properties of epidemics of plant pathogens have received scant attention to date. It is not at all apparent, however, that these properties should dictate the structure of models for disease progress.

Until recently, most attention in below-ground pathology was concentrated upon just two of these models, $\mathrm{d}Y/\mathrm{d}t = r(1 - Y)$, the monomolecular, and logistic, $\mathrm{d}Y/\mathrm{d}t = rY(1 - Y)$, models (see e.g. Gilligan, 1983). These models offered an appealingly simple classification of epidemics into just two types: the monomolecular model corresponded to monocyclic disease and the logistic to polycyclic disease. The curves for both models could easily be straightened by simple transformations of $\ln(1 - Y)^{-1}$ against t for the monomolecular and $\ln[Y/(1 - Y)]$ against t for the logistic model. Frequent use of these models has fostered their credibility amongst plant pathologists as useful descriptors of epidemics of soil-borne pathogens. This has happened more particularly since the myth that all soil-borne diseases are monocyclic was dispelled by Campbell and Powell (1980), Pfender (1982), and others.

There are, however, problems in the simplistic use of the logistic and monomolecular models. Morrall and Verma (1981) and Pfender (1982) drew attention to the logical flaw of making inferences about the nature of the disease cycle on the basis of goodness-of-fit tests to the models. Indeed, Pfender (1982) showed that it was possible to obtain a better fit to the monomolecular than to the logistic model for certain disease progress data of *Sclerotium cepivorum*, even though the disease was known to be polycyclic (Crowe and Hall, 1980). The model ought to be chosen on the basis of *a priori* knowledge of the biology of the system.

The nature of the disease variable is important. Expression of Y as proportion of diseased plants obscures secondary spread of lesions within plants (Morrall and Verma, 1981: Huisman, 1982). This bias is reduced by treating roots rather than plants as units of the host population. The use of discrete indices of disease, for example the use of a 0–5 scale, should be avoided when using non-linear models. The discrete nature of the variable may cause statistical difficulties and the distinction between increase in proportion of plants or roots infected is confounded with lesion growth (Morrall and Verma, 1981; Huisman, 1982).

Another problem with the standard non-linear models is the failure to take account of host growth. Huisman (1982) has discussed some of the biological consequences of host growth on disease progress. Some methods of including host growth in epidemiological models are considered in section II,C,1.

Even the apparent simplicity of the estimation of the rate parameters of the commonly used non-linear models is beguiling. In particular, the rate parameters, r, ought not to be estimated from the slope of a straight line fitted by ordinary least-squares regression to the appropriately transformed disease progress curve (Fulton, 1979). The variance of the disease variable is not constant over the entire range from 0 to 1. Alternative techniques such as maximum likelihood estimation or the use of weighted regression ought to be used (Ashton, 1972). Failure to do so renders the comparison of transformed curves biased. It is especially important that maximum likelihood procedures be used in the estimation of parameters of the three-parameter models such as the Weibull model (Pennypacker *et al.*, 1980). This has not always been done. Even with the two-parameter equations, correlation between the parameters may occur. Thus, Burdon and Chilvers (1975) report an increasing curvilinear response between the logistic rate of increase of damping-off of lettuce seedlings and initial density of inoculum of *Pythium irregulare*. Pfender and Hagedorn (1983) report a similar result for *Aphanomyces euteiches* f. sp. *pisi* on peas while Pullman and DeVay (1982) found that the arithmetic rate of increase of *Verticillium* wilt on cotton also varied with inoculum density. Given that initial levels of disease depend upon initial inoculum, there is evidence for correlation between the parameters of the simple two-parameter models of disease progress. It is interesting to note that negative rather than positive relationships have been reported for several aerial pathogens, including *Erysiphe graminis* on wheat (Rouse *et al.*, 1981), *Helminthosporium maydis* on maize (Gregory *et al.*, 1981) and *Uromyces phaseoli* on *Phaseolus vulgaris* (Plaut and Berger, 1981).

The supremacy of the monomolecular and logistic models in the description of epidemics has recently been challenged by the use of other models, notably the Gompertz (Berger, 1981) and the Weibull models (Pennypacker *et al.*, 1980). The Gompertz model, $\mathrm{d}Y/\mathrm{d}t = rY(-\ln Y)$, has the advantage of a non-symmetrical shape with more rapid increase in disease earlier than later. This has been shown to accord with observations made on epidemics of air-borne pathogens (Berger, 1981). It has yet to be rigorously tested for soil-borne pathogens. Being a three-parameter model, i.e. $\mathrm{d}Y/\mathrm{d}t = c[(t-a)/b]^{c-1}\exp[-(t-a)/b]^{c}$, the Weibull is more flexible than the Gompertz model. Biological meaning has been attributed to each of the three parameters: thus a is an estimate of the time of disease onset, b is an inverse function of the rate of development of the epidemic and c is the so-called shape parameter. Different values of c render the Weibull model approximately equivalent to simpler models such as the monomolecular ($c = 1.00$) and logistic ($c = 3.65$). Stack (1980) used the Weibull model in attempts to distinguish retrospectively between monocyclic and polycyclic progress of disease caused by *Helminthosporium sativum* on wheat. Campbell and Powell (1980) report a similar study for *Phytophthora parasitica* var. *nicotiana* on tobacco. Morrall and Verma (1981) challenged the use of a discrete disease index by Stack (1980), suggesting that this would not distinguish fairly between mono- and polycyclic disease. Campbell and Powell (1980) used percent diseased plants, but this was based on maximum numbers of plants per test plot of only 66. This is rather a small sample for unbiased estimation of population parameters. There is need for more detailed work, on the statistical properties of disease estimators used in the disease progress equations.

B. Linear Models

Although they are extremely flexible and have high predictive power, linear models have been used infrequently to describe the progress of soil-borne disease. They range in complexity from polynomial regressions to multiple regressions. Baker and Wijetunga (1981) found that a simple quadratic function could be used to describe the increase in inoculum density of *Rhizoctonia solani* under alfalfa. Last *et al.* (1969) used a multiple regression to describe the build-up of brown root rot of tomatoes, caused by *Pyrenochaeta lycopersici* within and between seasons. The model was of the form:

$$Y = B_0 + B_1 t + B_2 t^2 + B_3 S + B_4 S^2 + B_5 tS + \varepsilon$$

where Y is the proportion of diseased root, and t and S are the sampling times within and over seasons, respectively.

Unfortunately, there is one disadvantage that undermines the estimation of parameters of linear and indeed non-linear models. The amounts of disease present at successive times are not independent. They are in fact highly correlated and so too are likely to be the associated error terms. This being so, the usual technique of ordinary least-squares regression cannot be used to fit the equation to data. Various solutions to this problem of disease progress curves (DPCs) are discussed by Madden (1980). A general account is given in Neter and Wasserman (1976). Campbell *et al.* (1980b) used a first difference regression of the form

$$Y_t - Y_{t-1} = B_1(t_t - t_{t-1}) + B_2(t_t^2 - t_{t-1}^2) + (\varepsilon_t - \varepsilon_{t-1})$$

to describe the progress of hypocotyl rot of snap bean in 100 contiguous field plots. First difference regression is a relatively simple technique that involves regression of the change in disease on the change in time. The error terms $(\varepsilon_t - \varepsilon_{t-1})$ are more likely to be independent than the individual terms ε_t and ε_{t+1}. The procedure is therefore one of transformation so that ordinary least-squares regression can be used. Campbell *et al.* (1980a,b) investigated some of the relationships between parameters of first difference regressions and those of the Weibull model in describing epidemics of hypocotyl rot on *Phaseolus vulgaris*.

First difference regression is a simple example of a large family of autocorrelation techniques associated with time series analysis (Kendall, 1973b). Time series analysis is widely used in economic forecasting; its use in botanical epidemiology, especially with soil-borne pathogens, is restricted by the relatively small numbers of observations made on a single epidemic.

C. Custom-built Non-linear Models

A third option within the strategy of growth curve analysis is to foresake the standard models of population dynamics and to construct new, 'custom-built' models that satisfy assumptions appropriate to the system under study. The quest is for simple differential models that can be solved algebraically by integration, or failing that, by numerical methods. The parameters of these models should have some biological meaning, for if they do not the principal advantage of a non-linear over a linear model is lost.

Custom-built, non-linear models have so far been constructed for cyst nematodes (Jones *et al.*, 1967; Jones and Kempton, 1978; Jones and Perry, 1978) and for vesicular–arbuscular mycorrhizae (Buwalda *et al.*, 1982; Sanders and Sheikh, 1983; Smith and Walker, 1981; Tinker, 1975; Walker and Smith, 1984). Two principal objectives can be discerned in the construction of custom-built models for the population growth of symbionts. The first is to incorporate host growth into the models. This may be done either by the inclusion of empirical functions that require interpolation between experimental observations, or by the use of analytical functions such as the exponential function to approximate the growth of roots. The second objective is to distinguish between primary and secondary infection. This ordinarily requires the use of two equations rather than one.

I shall illustrate the construction of custom-built non-linear models by concentrating on vesicular–arbuscular mycorrhizal (v-a-m) fungi, in particular on the work of Buwalda *et al.* (1982), Smith and Walker (1981) and Walker and Smith (1984). The infection chain of v-a-m fungi is similar to that of ectotrophic root-infecting pathogens. Primary infection occurs from propagules in soil. An appressorium is formed, the root is penetrated and a short section of the cortex is colonised. External colonisation of the root surface occurs, from which it is believed that secondary infection of other roots and sections of the same root occurs (Buwalda *et al.*, 1982). The extent of internal infection from a single propagule is thought to be limited to a distance of approximately 1 cm (Buwalda *et al.*, 1982; Walker and Smith, 1984). This comprises an infection unit as defined by Cox and Sanders (1974).

1. Inclusion of Host Growth in Non-linear Models

Tinker (1975) drew attention to the false assumption in most non-linear models that the amount of host tissue available for infection of fungal symbionts was constant. Using total amounts rather than proportions of infected tissue, he proposed the following model for v-a-m fungi:

$$dL_i/dt = SL_i(nL_t - L_i) \tag{1}$$

where L_i and L_t are the lengths of infected and total roots, respectively, per unit volume of soil, and S is a constant of proportionality. The constant n was introduced because infection seldom reached 100%. Tinker (1975) actually used an analytical function, the exponential function, $L_t = A\ \exp(kt)$ to describe the change in root length over time. The model successfully fitted several sets of data for the infection of onions by a v-a-m fungus.

A simple heuristic justification of Eq. (1) is given by Buwalda *et al.* (1982). Suppose that most spread of infection occurs from external hyphae following primary infection from an undisclosed source. Then, assuming a completely random distribution of both roots and hyphae within a defined volume, the probability of new infections is proportional to $L_i \times L_t$. But additional infection on previously infected positions of root would not be detected, so L_t must be corrected to $L_t - L_i$. Whence $dL_i/dt = SL_i(L_t - L_i)$ or, for an asymptote less than one, $dL_i/dt = SL_i(nL_t - L_i)$.

Randomness of roots and of infected portions is an improbable assumption, especially since Buwalda *et al.* (1982) were considering experimental systems in which infection spread was from a single pad of inoculum in soil that was otherwise free of inoculum. Accordingly, they revised the model to account for spread as a 'front' moving along roots. The rate of formation of new infected root length (dL_i/dt) is therefore proportional to the number of fronts. The latter was unknown so a simple exponential relationship, $dL_1/dt = KL_i$, was chosen (Buwalda *et al.*, 1982). But some of this spread might be into already infected portions of root and so a correction term is employed. The probability that a small piece of root immediately adjacent to a 'front' is infected is L_i/L_t and so the probability of observing a new infection is $(1 - L_i/L_t)$. Note that this model implies that the fronts are randomly distributed relative to each other. The model for infection is given by

$$dL_i/dt = SL_i(1 - L_i/nL_t) \tag{2}$$

when the correction, n, is included. Eq. (2) is identical to the logistic model, save that L_t varies with time.

How can the models in Eqs. (1) and (2) be fitted to data? Tinker (1975) does not describe the method of fitting his model. The model cannot be readily integrated and so it is probable that it was fitted by numerical methods. This method was used by Buwalda *et al.* (1982). The objective in using a numerical method is to estimate values for L_i at successive intervals of time. Values for L_t were not estimated simultaneously from a model but were taken instead from observations of total root length per unit volume on the experiment to be modelled (Buwalda *et al.*, 1982). Intermediate values of L_t, i.e. between successive harvests of the experiment, were estimated by interpolation. Then, using a standard procedure for estimating the area under a curve known as the trapezoidal rule (see e.g., Causton, 1977), estimates of infected root length, L_i, are obtained as follows:

$$\hat{L}_i = \sum_{j=1}^{k} (t_j - t_{j-1})[(dL_i/dt_{t_j}) + (dL_i/dt_{t_{j-1}})]/2 \tag{3}$$

L_i is then replaced by $\hat{L}_i$ in the expansions of the differentials in Eq. (3) and the procedure repeated. Two or three iterations are usually sufficient to obtain an approximate solution that is satisfactory given the limits of experimental error (Buwalda *et al.*, 1982). Having obtained values for L_i over the time course of the experiment, a curve is fitted to these points using weighted regression and the parameters S and n estimated. The goodness of fit of the observed values of L_i with the models can be assessed visually.

The models were fitted to data for infection of spring wheat and white clover by *Glomus mosseae* in separate pot experiments (Buwalda *et al.*, 1982). Inoculum was placed below the central plant in each pot. Both models fitted the data reasonably well, but that given in Eq. (2) was always superior at low values of L_i. Neither host density nor host species affected S but the estimate of the correction factor n was much greater for clover than for wheat. The authors concluded that differences in susceptibility to infection between species should be sought not in the factors allowing initial rapid extension but in the unknown mechanisms which control the final extent of infection which they presume resides in the host. In judging the goodness of fit of the model it should be noted that the model does not include infection spreading from one plant to another but this is known to occur.

2. *Further Refinements to Non-linear Models*

Tinker (1975) and Buwalda *et al.* (1982) were concerned with the inclusion of host growth in a model for the progress of infection but neither model distinguished between primary and secondary infection. Smith and Walker (1981) and Walker and Smith (1984) constructed models that incorporated host growth and made partial distinction between primary and secondary infection by modelling changes in the number of infections and in the length of infected root separately. They recognised three processes in the dynamics of root colonisation, root growth, the occurrence of infection, and fungal growth (Smith and Walker, 1981) and proposed two equations to describe the system:

$$\mathrm{d}U/\mathrm{d}t = A(L_t - L_i) \tag{4}$$

$$\mathrm{d}L_i/\mathrm{d}t = BU(1 - L_i/L_t) \tag{5}$$

where U is the number of entry points (or infection units) per plant, A is the frequency of infections per unit length of root per unit time, L_t and L_i are defined as above, save that Smith and Walker (1981) express them on a per plant basis, and B is the rate of spread of mycelium within the root from a single infection point. Thus $B/2$ is the rate of linear growth of mycelium within the cortex. New infections were assumed to arise only in uninfected portions of the root, hence the use of the correction term $(L_t - L_i)$ in Eq. (4). This assumption may lead to under-estimation of U and of A and B. The correction term in Eq. (4) was chosen empirically for its simplicity (Walker and Smith, 1984). Further important assumptions of the models are that uninfected portions of roots are uniformly susceptible to infection, and that the rate of growth of the fungus is constant over the entire root system.

Smith and Walker (1981) fitted their model to data for the infection of *Trifolium subterraneum* by *G. mosseae* in pot experiments. Like Buwalda *et al.* (1982), they used observed values for L_t and numerical methods of integration to obtain estimates for U and L_i. Separate estimates of the parameters A and B were obtained for successive intervals of time. The authors then examined the trend in A and B over time, finding that the parameters were constant over the period 5–35 days after planting. In addition, the parameters differed for main and for lateral roots and both parameters were influenced by changes in the density of inoculum.

The original variables, U, L_i and L_t were expressed on a per plant basis and the means of groups of plants were used in testing the models. But L_t varies from plant to plant, and this in turn is likely to produce correlated variation in U and L_i (Walker and Smith, 1984). One possible consequence of using U, L_i and L_t, therefore, is that variances of experimental results are likely to be high, rendering the experiment and the models imprecise. Walker and Smith (1984) suggest a simple alternative of expressing the dependent random variables on the basis of per unit length of root rather than per plant, in order to reduce the variance inherent in the original models. The new variables are l, the proportion of the length of root infected ($l = L_i/L_t$), and u, the frequency of infection units per unit length of root ($u = U/L_t$).

Consider the transformation of Eq. (4) using the expansion for the differentiation of a product:

$$\mathrm{d}(uL_t)\mathrm{d}t = L_t(\mathrm{d}u/dt) + u(\mathrm{d}L_t/\mathrm{d}t) = A(L_t - L_i)$$

$$\mathrm{d}u/\mathrm{d}t = A(L_t - L_i)/L_t - u(\mathrm{d}L_t/\mathrm{d}t)/L_t$$

Let $g = (\mathrm{d}L_t/\mathrm{d}t)L_t^{-1}$, the relative growth rate of roots, thus

$$\mathrm{d}u/\mathrm{d}t = A(1 - l) - gu \qquad (6)$$

Similarly, transformation of the variable in Eq. (5) gives

$$\mathrm{d}l/\mathrm{d}t = BU(1 - l) - gl \qquad (7)$$

Eq. (6) and (7) can be solved numerically: the objective is to estimate the parameters A and B between successive harvest times, t_{j-1} and t_j. The value of A within a given interval is designated as A^j_{j-1} and the corresponding value for B is B^j_{j-1}. The variables $(L_t)_j$, l_j and u_j are measured at each of j harvest times. Estimates for L_t, $\mathrm{d}L_t/\mathrm{d}t$, l, $\mathrm{d}l/\mathrm{d}t$, u and $\mathrm{d}u/\mathrm{d}t$, are required in the interval t_{j-1} and t_j. Walker and Smith (1984) do this simply by interpolation, thus:

$$L^j_{t(j-1)} = [L_{t(j)} + L_{t(j-1)}]/2$$

$$\left(\frac{\mathrm{d}L_t}{\mathrm{d}t}\right)^j_{j-1} = [L_{t(j)} - L_{t(j-1)}]/(t_j - t_{j-1})$$

Walker and Smith (1984) obtained good agreement between the models and observations of pot experiments of *G. mosseae* on *T. subterraneum* over periods of 35 days. Several trends were apparent from the results. For example the rate of initiation of new infections, A, was directly related to propagule density at early harvests; the rate of growth of infection units, B, was unaffected by propagule density, but declined with time. The relationship between A and propagule density was obscured at later harvests by, it was presumed, interactions between primary and secondary infections. The authors speculate that the fall in B may be related to the age structure of the population of infection units, i.e. the growth rate slows as individual infection units become older. It might also be correlated with the age of the plant.

The models in Eqs. (6) and (7) can be solved numerically and the trends in the values of the two parameters, A and B, used to make inferences about the underlying epidemiological mechanisms. Walker and Smith (1984) also derive some explicit solutions, under certain limiting assumptions, for short times and for long times. Consider first the very short term $u = 0, l = 0, t = 0$, when $\mathrm{d}u/\mathrm{d}t \approx A$ and $\mathrm{d}l/\mathrm{d}t \approx Bu$. Integration and substitution give $l = BAt^2/2$ and $u = At$, where $l/u = Bt/2$, i.e. the mean length of an infection unit is initially independent of A.

Of more interest, perhaps, are the limiting values of l and u at long times. If l and u are asymptotic, $\mathrm{d}l/\mathrm{d}t$ and $\mathrm{d}u/\mathrm{d}t$ will be zero. The derivation of an expression for l_∞, the asymptotic value of l, is given below to illustrate the procedure.

Let

$$0 = A(1 - l_\infty) - gu \qquad \text{and} \qquad 0 = Bu(1 - l_\infty) - gl_\infty$$

$$u = A(1 - l_\infty)/g$$

So,

$$0 = AB(1 - l_\infty)^2/g - gl_\infty, 0 = ABl_\infty^2(2AB + g^2) + AB$$

There are two solutions to this quadratic equation:

$$\begin{aligned} &1.\ l_\infty = [(4AB + g^2)^{1/2} + g]/[(4AB + g^2)^{1/2} - g] \\ &2.\ l_\infty = [(4AB + g^2)^{1/2} - g]/[(4AB + g^2)^{1/2} + g] \end{aligned} \tag{8}$$

The first solution is rejected because l_∞ cannot be greater than one. It may be seen from Eq. (8) that l_∞ must of necessity be less than 1.0. This result led Walker and Smith (1984) to reject the validity of the correction term proposed by Tinker (1975) and Buwalda *et al.* (1982). Similarly an expression may be derived for u_∞:

$$u_\infty = [(4AB + g^2)^{1/2} - g]/2B \tag{9}$$

Dividing Eq. (8) by Eq. (9) gives an expression for the apparent mean length of an infection unit, m_∞:

$$m_\infty = 2B/[(4AB + g^2)^{1/2} + g] \tag{10}$$

Note that the constancy of the expressions in Eqs. (8) to (10) implies a constant relative growth rate, g, for the roots.

The model of Walker and Smith (1984) is, heuristically at least, more realistic than that of Tinker (1975) and Buwalda *et al.* (1982). Walker and Smith (1984) use two parameters that describe the apparent rate of increase in the number of infections and the mean rate of growth of infections. The formation of new infection units and fungal growth are inseparable in the models of Tinker (1975) and Buwalda *et al.* (1982). Walker and Smith's model has the advantage of applying to situations in which inoculum is randomly dispersed in soil. The model does not, however, permit clear distinction between primary and secondary infection.

3. *Analytical Functions for Host Growth in Non-linear Models*

We have seen that an analytical function for host growth might be included in a non-linear model. Thus Tinker (1975) used an exponential function to describe the change of L_t with time. Turner *et al.* (1969) discuss the use of a logistic model to describe the change of the asymptote up to an ultimate population maximum, i.e.

$$dL_i/dt = S'L_i[k(t) - L_i]$$

in which $k(t)$ is itself a logistic function.

Turner *et al.* (1969) give an analytical solution to their model so that, in this case, L_i might be obtained as a function of L_t and time. Analytical functions of combinations of non-linear models, however, are difficult to solve. Kushalappa and Ludwig (1982) discuss ways of obtaining estimates for the rate parameter of logistic and other models in which the amount of host tissue changes over time. They also fit disease progress data to these functions although the method of fitting is not clearly described.

Quite apart from the intricacies of fitting the models to data, the inclusion of asymptotic functions for host growth in models for disease progress involves two important assumptions. The first is that there is an asymptote which is known. The second is that host growth is unaffected by the presence of the parasite. Buwalda *et al.* (1982), Smith and Walker (1981) and Walker and Smith (1984) avoided this problem by using observed values rather than an analytical function for the amount of host available for infection at each harvest. Nematologists, however, have considered the problem of using analytical functions to describe the effects of infection on yield (Seinhorst, 1965; Jones and Kempton, 1978; Jones and Perry, 1978). Seinhorst (1965) argued that the yield of plants was unaffected when nematode density at planting or sowing was below a certain threshold *T.* Above this threshold, yield decreased exponentially with egg density; thus, using the notation from above and the form of the equation given by Jones and Perry (1978),

$$L_t/L_{t(\max)} = L_{t(\min)}/L_{t(\max)} + (1 - L_{t(\min)}/L_{t(\max)})z^{P_i - T} \tag{11}$$

Where L_t is the expected yield, $L_{t(\max)}$ the yield in the absence of nematodes, and $L_{t(\min)}$ the yield given a very large initial nematode density, P_i is the initial density of eggs and z a constant. Jones and Perry (1978) cite evidence (Lownsbery and Peters, 1955; Jones 1957; Peters, 1961) that there is some compensation by plants at small nematode densities so that yields may exceed those in uninfested soil. Therefore, they argued that Eq. (11) held for all values of P_i. Taking $L_{t(\min)}$ to be zero for roots and z^{-T}, equal to c, say, a constant slightly greater than one, to account for compensation, Jones and Perry (1978) revised Eq. (11) to

$$L_t = L_{t(\max)} c z^{P_i}$$

This function was incorporated into a simple difference model in which the number of juveniles per centimetre root was estimated from the initial egg density but with due allowance for host–parasite interaction. Although the model involves a simple system of a parasite with discrete non-overlapping generations, it does reveal one way in which a negative feedback of infection on host growth might be incorporated into models for the population dynamics of plant pathogens. More quantitative studies on this aspect of modelling are needed.

III. MODELS FOR COMPONENTS OF THE INFECTION CHAIN

A. Primary Infection

Models for primary infection have two specific objectives: to estimate the probability that an infection court, such as a root, will be infected given certain densities of roots and inoculum; to estimate the mean or maximum distance from the surface of a host from which infection can be initiated. A third, more general, objective is to analyse and synthesise the dynamic components of this part of the infection chain. Dynamic components comprise such factors as efficiency of inoculum for infection, rate of growth of the host and rate of change of susceptibility of sites for infection (Gilligan, 1985b).

Two broad approaches are to be found in models for primary infection (Gilligan, 1983): one is based on elementary probability theory, the other upon theories of solid packing and surface density, as developed, for example, in physical chemistry (Baker, 1971; Baker and Drury, 1981; Gyani, 1945). I propose to restrict the discussion to probability models. Recent discussion of the surface density models is to be found in Drury *et al.* (1983) and Ferriss (1983).

Primary infection is the result of a sequence of chance events (Gilligan, 1985b). The sequence is initiated by the occurrence of a propagule close enough to a host to have some chance of infecting it; that is, the propagule lies within the pathozone (*sensu* Gilligan, 1985a), which is the region of soil surrounding a host unit within which the centre of a propagule must lie for infection of the host unit to be possible. Occurrence in the pathozone is followed by growth or, if the propagule is mobile, movement to the surface of the host, colonisation and infection. For simplicity, it is convenient initially to consider a static host such as a seed or an infection site behind a growing root tip.

We consider first the construction of models for the probability of host infection for both random and non-random distribution of inoculum. Models for changes in propagule numbers are then discussed.

1. Modelling the Probability of Host Infection: Stationary Infection Courts

Consider a volume V of soil in which there are P propagules and N seeds. Suppose that a propagule can only cause infection if it occurs within a certain critical distance from the surface of a seed, i.e. within the pathozone. The probability that any one propagule occurs within a pathozone is given by $\phi = v/V$ where v is the volume of the pathozone. Now, proceeding deterministically, the total number of encounters, and hence infections per seed, say N_e/N, will be given by:

$$N_e/N = P\phi \tag{12}$$

Rearrangement of the model permits estimation of ϕ and hence also the dimensions of v:

$$\phi = N_e/NP$$

This simple model was proposed by Gilligan (1979, 1980a) and amended by Ferriss (1981). It had the disadvantage that the number of infections had to be observed by laborious and difficult counting of separate infections or else these had to be estimated from the percent infected roots by the multiple infection transformation. Given the assumption of random occurrence of propagules, it is simple matter to reformulate the model in the form of a binomial model. For one infection court and one propagule:

$$\text{Prob. [infection court remains uninfected]} = (1 - \phi).$$

So the total number of hosts that are uninfected, N_u say, is given by:

$$N_u = N(1 - \phi)^P \tag{13}$$

Rearrangement allows solution for ϕ:

$$\phi = 1 - (N_u/N)^{1/P} \tag{14}$$

The problem of multiple infection is automatically accounted for in the model. It is necessary only to score for the proportion of uninfected infection courts.

If v is small, i.e. $v \ll V$, and P large, the model may be rephrased using the first term of the Poisson rather than of the binomial distribution to estimate the probability that a seed escapes infection. Thus,

$$N_u = N \exp(-\phi P) \tag{15}$$

where ϕP is the mean number of propagules per pathozone (cf. the derivation of the one-hit, Poisson equation given by Van der Plank, 1975, p. 24). Rearrangement gives

$$\phi = v/V = [\ln (N/N_u)]/P \tag{16}$$

Estimation of the radius of v from Eqs. (14) and (16) is not, however straightforward. Two models for the shape of the pathozone have been proposed (Ferriss, 1981; Gilligan, 1979, 1985a), namely a displacement model in which the volume of the pathozone is considered to be solid, and a shell model for which the volume is hollow. It is probable that the shell model applies to seeds. Thus v is a hollow sphere surrounding the seed and

$$v = 4\pi(R^3 - r_{sd}^3)/3$$

where R is the outer radius of the pathozone and r_{sd} the radius of the infection court, in this case the seed. With larger targets such as some roots, growth of the host involves displacement of the soil and of the propagules within it. If the furthest extent of the change in bulk density is less than $R - r_{sd}$ from the root surface, then the displacement model holds. Thus for a root of length L, uniformly susceptible over its surface, $v = \pi LR^2$. For further discussion of these models, see Ferriss (1981) and Gilligan (1985a).

The assumptions common to the binomial and Poisson models are twofold: infection occurs whenever a propagule occurs within a pathozone and the occurrence of propagules within a pathozone is random. These assumptions are severe and unrealistic. Not all propagules are necessarily capable of causing infection (Baker and Drury, 1981) and the probability of infection declines with increasing distance of infective propagules from the host (Henis and Ben-Yephet, 1970; Gilligan, 1980a; Punja and Grogan, 1981). The spatial pattern of inoculum is more likely to be aggregated than random (Gilligan, 1983). Significant deviation from randomness has been demonstrated for inoculum of *Cylindrocladium crotalariae* in bulk soil (Taylor *et al.*, 1981; Hau *et al.*, 1982) and for inoculum of *Pythium aphanidermatum* in bulk soil (Stanghellini *et al.*, 1982) and in the pathozone of sugar beet (Stanghellini *et al.*, 1983). We consider now how the models can be adjusted to include more realistic assumptions. We shall assume random distribution of inoculum initially and consider the revised statement (Gilligan, 1985a):

$$\text{Prob. [infection court remains uninfected]} = (1 - \theta\psi)$$

in which θ is the probability of occurrence of a propagule within a pathozone and ψ is the probability of a propagule infecting the host, conditional upon its occurrence in the pathozone. Assume a shell model for infection of a seed. Then:

$$\psi = \int_{r_{\text{sd}}}^{R} (\psi_1|\theta)(\psi_2)\mathrm{d}r$$

where $(\psi_1|\theta)$ is the density function for the probability that the centre of a propagule occurs between say r and $r + \Delta r$ from the centre of the pathozone, conditional upon θ, the occurrence of the propagule within the pathozone; (ψ_2) is the density function for the probability that the propagule can infect the host, conditional upon its occurrence between r and $r + \Delta r$. For a seed,

$$\psi_1 = \frac{4\pi r^3 \Delta r/3}{\sum_{r_{\text{sd}}}^{R} 4\pi r^3 \Delta r/3} = \frac{r^3}{R^3 - r_{\text{sd}}^3}.$$

A simple expression may be selected for ψ_2 (Gilligan, 1985a), for example:

$$\psi_2 = \beta \exp(-\alpha r^2)$$

Whence,

$$\psi = \int_{r_{\text{sd}}}^{R} \frac{\beta r^3}{(R^3 - r_{\text{sd}}^3)} \exp(-\alpha r^2)\mathrm{d}r$$

with a solution (Gilligan, 1985a):

$$\psi = \frac{\beta}{2\alpha^2(R^3 - r_{\text{sd}}^3)} [\exp(-\alpha r_{\text{sd}}^2)(1 + \alpha r_{\text{sd}}^2) - \exp(-\alpha R^2)(1 + \alpha R^2)]$$

In practice, θ and ψ will be affected by the relative growth of roots, the number and duration of susceptible sites on the roots (Huisman, 1982; Ferriss, 1982, 1983) and the rate of germination and growth of pathogen propagules. Suppose now that propagules are not randomly distributed but are aggregated. The negative binomial distribution may be used to approximate the spatial pattern of the propagules in soil (Gilligan, 1983). The first term of the distribution, $(1 + \phi P/k)^{-k}$ in which ϕP is the mean density of propagules within the pathozene and k is an index of aggregation, gives the probability of zero occurrence. Substitution into Eq. (13) gives

$$N_u = N(1 + \phi P/k)^{-k} \tag{17}$$

It is probable that more than one infection may be necessary to cause visible symptoms of disease. The models given above can be adapted to take account of numerical thresholds of infection. Suppose for example that death of a seed occurred only after $> j$ independent infections. Equations for the expected numbers of dead hosts (N_d) are as follow:

binomial model:

$$N_d = N\left[1 - \sum_{n=0}^{n=j} \binom{P}{n} \phi^n (1 - \phi)^{P-n}\right] \tag{18}$$

Poisson model:

$$N_d = N\left[1 - \sum_{n=0}^{n=j} (\phi P)^n \exp(-\phi P)/n!\right] \tag{19}$$

Negative binomial model:

$$N_d = N\left[1 - \sum_{n=0}^{n=j} \binom{n+k-1}{k-1} (\phi P/k)^n (1 + \phi P/k)^{-(k+n)}\right] \tag{20}$$

The second term within the square brackets in Eqs. (18) to (20) gives the probability that the host will be alive. The effect of random and nonrandom ($k = 0.5$) occurrence of propagules on the relationship between propagule density and the probability of escape from disease is shown in Fig. 2, after Gilligan (1985a), with j arbitrarily set to 3. Clumping of inoculum results in proportionately fewer diseased hosts than randomly dispersed inoculum with increasing inoculum density.

2. Modelling Changes in Propagule Numbers

So far I have been concerned principally with the use of probability models to estimate the probability of escape from infection. It is apparent, however, from Eq. (12), that it is a simple matter to estimate the numbers of encounters or infections when host and inoculum density are known, at least when encounters occur at random. To estimate the change in propagule numbers, some allowance for density dependence in the relationship $N_e = NP\phi$ is

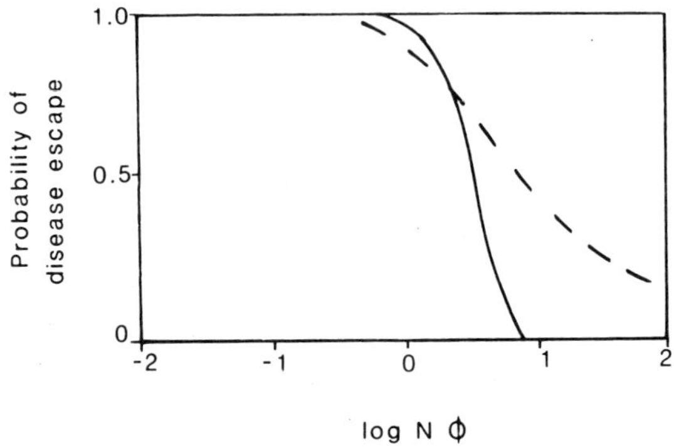

Fig. 2. Relationships between inoculum density, $N\phi$, within a hypothetical pathozone for the binomial model (—) for random distribution of inoculum, and the negative binomial model (- -) for aggregated distribution of inoculum. The index of aggregation was set at $k = 0.5$ and hosts were assumed to escape disease if they encountered $\leq$ three propagules of inoculum.

necessary. Jones and his co-workers [Jones *et al.* (1967), Jones and Kempton (1978) and Jones and Perry (1978)] provide one solution. It is derived for a population of cyst nematodes such as *G. rostochiensis*, an organism which is monocyclic within a season.

Suppose that P_i and P_f are the initial and final numbers of eggs in unit volume of soil during a season. Of the P_i eggs, a certain proportion C do not hatch and of those that do a certain proportion H succeed in invading the roots, of total length h cm, of the host crop. The proportion H is assumed to be independent of density, at least up to a certain density (Jones and Kempton, 1978). The number of larvae per centimetre root, P_r, is given by

$$P_r = H(1 - C)P_i/h \tag{21}$$

There are two further stages at which density-dependent regulation might be expected to operate. Either it might occur in the ratio of females to males (Trudgill, 1967) or in the numbers of eggs laid per female. Jones and Perry (1978) suggest that, in the absence of predators and parasites, the main density-dependent regulatory mechanism appears to be the regulation of the sex ratio. They note that the ratio of males (M) to females (F) increases proportionately with the numbers of juveniles per centimetre root. Thus,

$$M/F = bP_r \tag{22}$$

where b is a constant; and

$$M + F = P_r \tag{23}$$

Recombining Eqs. (21), (22) and (23) gives a logistic model for the numbers of females in roots:

$$F = [H(1 - C)P_i/h][1 + bH(1 - C)P_i/h]$$

The number of females per gram soil is

$$F' = H(1 - C)P_i[1 + bH(1 - C)P_i/h]$$

Suppose now that each female produces an average of m' eggs and that this is independent of density (Jones and Kempton, 1978), then

$$P_f = m'F' + CP_i \tag{24}$$

Allowance for the interaction between parasite and host growth, as discussed above, was included in the model (Jones and Perry, 1978) by replacing h by hcz^{P_i} in the above equations. Note that Jones and Perry (1978) and Jones and Kempton (1978) recast their equations so that population densities were expressed as proportions of the equilibrium densities, i.e. the carrying capacities, with and without root damage.

The models were used for both theoretical and practical purposes. Jones and Kempton (1978) investigated the effects of changes in the population density in the region of the carrying capacity. They also fitted a computer model, based on Eq. (24), successfully to data for *G. rostochiensis* obtained from field plots in which susceptible varieties of potato were grown for 5 or more successive years. Various extensions to the models were considered (Jones and Perry, 1978). These include the effects of growing resistant crops, of integrated control measures and of parasitism of eggs by fungi. A detailed discussion of the use of probability models to describe the effects of parasitic fungi on the population dynamics of *Heterodera avenae* is given by Perry (1978). For alternative models for nematode increase, see Seinhorst (1970).

The models discussed above are of course specific to the infection chains of monocyclic nematodes. They do however illustrate how probability models may be used to model changes in populations density. They also suggest avenues for research with pathogenic fungi. The density-dependent effect of M versus F in the cyst nematodes might be transposed to nonsporing versus sporing infections or density dependence may for example operate upon m' the average number of offspring per sporulating infection. Little thought has so far been given to regulatory mechanisms in the population dynamics of soil-borne pathogens.

3. Relationships between Probability Models and Nicholson–Bailey Models

There is a marked similarity between the probability models for a stationary infection court and the Nicholson–Bailey models for host–predator and host–parasitoid behaviour (Nicholson, 1933; Nicholson and Bailey, 1935).

The Nicholson–Bailey models rest on two assumptions. The first is that the number of encounters with hosts, N_e, by say P parasitoids is in direct proportion to host density, hence:

$$N_e = aNP \tag{25}$$

where a is a constant. The second is that searching by the predator is such that these N_e encounters are distributed randomly amongst the available hosts with mean N_e/N. The number of hosts that escape attack (N_u) is therefore given by

$$N_u = N\exp(-N_e/N) = N\exp(-aP) \tag{26}$$

The number that are attacked, e.g. parasitised, N_p is given by

$$N_p = N[1 - \exp(-aP)] \tag{27}$$

Equations (25), (26) and (27) are identical to eqs. (12), (15) and (19) (the latter with $n = 0$). Hence the parameter a described by Nicholson (1933) as the 'area of discovery' and familiar to entomologists as the 'searching efficiency' of a parasitoid or predator is similar to ϕ, the probability that a propagule occurs in a pathozone and infects the host. Equation (27) describes the 'Nicholson competition curve', more familiar to plant pathologists as the inoculum density-disease incidence curve (Fig. 3).

The models of Nicholson and Bailey (1935) were proposed to describe the changes in population size in successive non-overlapping generations. Thus for a host with a uniform rate of increase, λ,

$$N_{t+1} = \lambda N_u = \lambda N_t \exp(-aP_t) \tag{28}$$

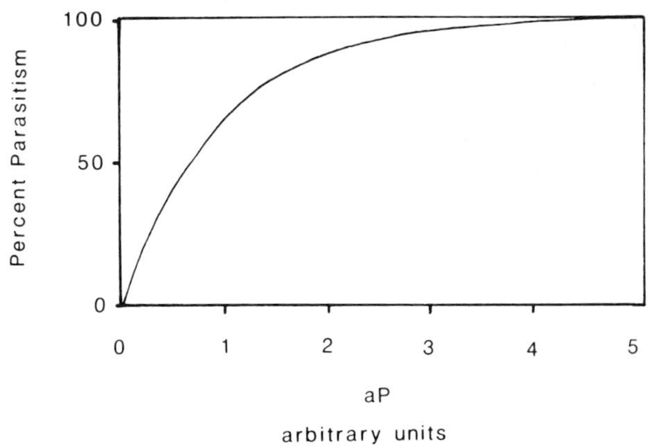

Fig. 3. The Nicholson–Bailey competition curve in which percent parasitism is estimated by the expression $1 - \exp(-aP)$ where a is the searching efficiency and P is parasite density. The curve is analogous to an inoculum density–disease incidence curve when one infection can cause disease and aP is the mean density of inoculum within a pathozone.

and for a parasitoid for which an infested host supports a single larva,

$$P_{t+1} = N_p = N_t[1 - \exp(-aP_t)] \tag{29}$$

Theoretical and practical work in entomology has been directed at the stability behaviour of Eqs. (28) and (29) (see e.g. Hassell, 1976, 1978). Of greater immediate interest to plant pathologists is the way in which variability in the searching efficiency, *a*, can be modelled. It was assumed in the derivation of Eq. (25) that *a* was independent of prey and predator or parasitoid density. A similar assumption, that ϕ is independent of host and pathogen density, is implicit in Eq. (12). These assumptions are unlikely to hold for many cases.

The situation of predators is not of course directly analogous to that of plant pathogens. That of parasitoids, in which the host remains after attack, is closer, but still most fungal parasites multiply almost continuously. Consider, however, a population of a damping-off fungus exposed to seedlings in soil. There is a fixed period during which the hosts are susceptible. Moreover, it is possible to relate the performance of a parasite to the initial inoculum density rather than to the instantaneous density of inoculum or biomass. The initial inoculum therefore is analogous to the population of parasitoids and the seedling density to prey density.

Consider changes in the host density. The total time available for attack can be divided into two components, T_s, the time available for 'searching', and T_h, the time spent in 'handling' (i.e. colonising) the host. Now the searching efficiency can be divided into two components such that $a = a'T_s$ where a' is the instantaneous search rate (Hassell, 1978). The units for *a* are encounters per host per parasite and those for a' are encounters per host per parasite per unit time.

Holling (1959) proposed three types of relationship between prey density and the number of prey attacked per predator (Fig. 4). These were called functional responses after Solomon (1949). A functional response is a change in the numbers of prey attacked in a fixed period of time by a single predator or parasitoid when the initial prey density is changed. (The abscissa of the functional response is conventionally shown as N_a, the number of attacks. In this discussion, in which random rather than systematic search is considered, it can more properly be thought of as the number of encounters per parasite, N_e/P.) For a type I response (Fig. 4A), $T_h \ll T$ below a certain threshold, say N_x, so $a'T_s$ is a constant and

$$N_e/P = a'T_sN \tag{30}$$

If search is random (Rogers, 1972),

$$N_p = N[1 - \exp(-a'T_sP)] \qquad N < N_x$$

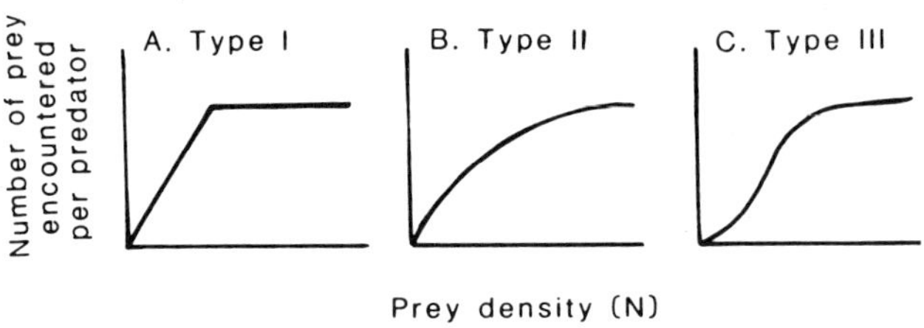

Fig. 4. Three types of functional response, after Holling (1959).

Above the threshold, it is assumed that there is no further attack. The type II response (Fig. 4B) is thought to be common amongst insect populations (Hassell, 1976) where handling time is significant. Thus,

$$T_s = T - T_h(N_e/P)$$

and substituting for T_s in Eq. (30) gives

$$N_e/P = a'TN(1 + a'T_hN) \tag{31}$$

whence,

$$N_p = N\{1 - \exp[-a'TP(1 + a'T_hN)]\}$$

Equations to describe the type III response have been proposed (Hassell *et al.*, 1977) but these rely heavily on inferences about the behaviour of insects. In principle, however, Hassell (1978) suggests that sigmoid responses might result from a', T_s or both being increasing functions of prey density. Hassell *et al.* (1977) propose a model in which a' is the variable parameter. Thus if

$$a' = bN/(1 + cN)$$

where b and c are constants then substituting in Eq. (31) gives

$$N_e/P = bN^2T_s/(1 + cN + bT_hN^2)$$

and, for a parasitoid,

$$N_p = N\{1 - \exp[-bTNP/(1 + cN + bt_hN^2)]\}$$

The relevance of these models to soil-borne plant pathogens is undoubtedly speculative. There have been few studies on the relationships between host density and relative success of inoculum in infecting hosts, but see Burdon and Chilvers (1975).

The searching efficiency may also vary with P. Entomologists have used an empirical relationship to relate a to predator density,

$$a = QP^{-m} \tag{32}$$

where Q and m are constants. The parameter m is a measure of interference amongst predators and the Nicholson–Bailey model is a special case when $m = 0$. Substituting Eq. (32) into Eq. (27) gives

$$N_{\mathrm{p}} = N[1 - \exp(-QP^{1-m})]$$

More elaborate models have been constructed in which wasted time due to mutual interference is introduced into the models. The effects of interference amongst propagules of inoculum of plant pathogens has, as yet, received little study. Henis and Ben-Yephet (1970) and Gilligan (1980b) have examined the effects of competition between propagules of inoculum at sites of infection but only in relation to the colonisation of a single host.

The models discussed above all assume a random search. Examination of the entomological literature shows that non-random search has been incorporated into the Nicholson–Bailey models. Hassell and May (1973) use an empirical method to incorporate aggregation of the predator and prey into n patches such that

$$N_{\mathrm{p}} = N\left[1 - \sum_{i=1}^{n} \alpha_i \exp(-a\beta_i P)\right]$$

where α_i and β_i are the fractions of the prey and predator, respectively, in the ith patch such that the sum of the α_i and β_i are each equal to unity. May (1978) uses the negative binomial distribution to describe aggregation of the predator in an analogous way to that described in section III,A,1.

B. Secondary Infection

Two sorts of secondary infection can, in principle, be distinguished (Gilligan, 1985b), reinfection of the same infection court and cross-infection to another infection court. The former, autoinfection *sensu* Robinson (1969), is frequently achieved by growth of mycelium and concomitant expansion of lesions. It includes growth up one root and down another via the crown as happens, for example, in *Gaeumannomyces graminis*. It might also be taken to include the vascular colonisation of hosts by wilt-inducing fungi such as *Verticillium* spp. Secondary infection between infection courts, alloinfection *sensu* Robinson (1969), involves infection of, say, one root from another because the roots happen to be close together rather than because they are anatomically connected.

Little has yet been published on secondary infection. I propose to confine discussion to the ectotrophically spreading fungi such as *G. graminis* and the vesicular–arbuscular mycorrhizal fungi such as *G. mosseae*.

I have shown that the custom-built non-linear models make some distinctions between primary and secondary infection, by modelling the advance of infection fronts when primary infection is all but completely eliminated (Buwalda *et al.*, 1982) and by distinguishing the change in length of infected root from change in numbers of infections (Smith and Walker, 1981). More subtle distinctions could be made by writing separate equations for the change in numbers of primary and secondary infections. Sanders and Sheikh (1983) describe models for the growth of infection units of *Glomus caledonius* and for the spread of infection in root systems. Although these models provide insight into the dynamics of the system, they are arguably more descriptive than analytical. They assume randomness of infection and non-interference between infections, other than by overlap. The models therefore do not address the problems of the interaction between the spatial pattern of roots and of infections, [i.e. the occurrence of infections relative to each other and the geometry of host growth (Gilligan, 1982b)].

Of central importance in introducing the spatial pattern of root growth into models for secondary infection is the need to estimate the probability of cross-infection between infection courts. One approach to this is to consider the frequency distributions of distances between infection courts. Only when, say, roots come closer than a certain critical distance can cross-infection be expected to occur. The problem is analogous to that of a propagule occurring within a pathozone, although the dynamics of the system are more complex. Data on the distributions of distances amongst plant roots are scarce but Barley (1970) notes that efficient methods are available for the measurement of rooting density and pattern on a fine scale. Originally, the ends of roots protruding from the faces of soil cores were counted (Fitzpatrick and Rose, 1936). Later polished surfaces and low-power microscopy were used (Barley and Sedgeley, 1961; Melhuish, 1968). Now most of the techniques involve autoradiography and sectioning of soil cores, so that the positions of roots can be mapped as point processes on a plane surface (Baldwin *et al.*, 1971). Fusseder (1983) examined the pattern of roots of *Zea mays* in serial sections of maize. He found frequent deviations from randomness towards aggregation of the numbers of cross-sections of roots per unit surface area. Fusseder (1983) used the variance/mean ratio to describe the system but nearest neighbour methods of analysis (see e.g. Pielou, 1977) might also be used to test for deviation from randomness independently of scale and to summarise the frequency distribution of distances between roots. Partial excavation has been used to estimate the number of root contacts or near contacts, especially in forest studies (Reynolds and Bloomberg, 1982). Thus Reynolds and Bloomberg (1982), working with *Phellinus weirii* on Douglas fir, used the multivariate statistical technique of stepwise discriminant analysis to determine the influence of stand and site factors on the probability of root contact.

Difficulty lies in considering inter-root distances over the entire lengths of roots. Instead of seeking an analytical solution, Gilligan, Campbell and Brassett (unpublished data) are using a mechanistic computer simulator to describe the paths of growth of the main axes of seminal roots of wheat that are exposed to inoculum of *G. graminis*. Knowing the geometry and rates of growth of roots, it is possible to estimate the number of primary infections, the extent of reinfection of roots and the number and extent of cross-infections by simulation. The first seminal root is assumed to grow vertically down, the rest are approximated by the equations for three-dimensional parabolas, with variable angles of rotation (α) from the crown. Thus for a root whose crown is at the point $(x_0, y_0, 0)$ the subsequent position of any point (x,y,z) on the root may be expressed in the form $(x_0 + 2pt \cos \alpha, y_0 + 2pt \sin \alpha, pt^2)$ where the parameter p determines the shape of the parabola and t is the parametric value of the point (x,y,z). The parametric coordinates satisfy the equation for a three-dimensional parabola:

$$(x - x_0)^2 + (y - y_0)^2 = 4pz$$

The minimum distance between an infected point (x,y,z) to another root is calculated as follows. The distance δ between (x,y,z) on the infected root and any point t on the uninfected root is given by:

$$\delta^2 = (x - x_0 - 2pt \cos \alpha)^2 + (y - y_0 - 2pt \sin \alpha)^2 + (z - pt^2)^2 \quad (33)$$

The parameter t which represents the point on the uninfected root nearest to (x,y,z) is determined by minimizing δ^2 with respect to t, i.e. $d(\delta^2)/dt = 0$.

So,

$$2(x - x_0 - 2pt \cos \alpha)(-2p \cos \alpha) + 2(y - y_0 - 2pt \sin \alpha) \times (-2p \sin \alpha) + 2(z - pt^2)(-2pt) = 0$$

And,

$$p(x - x_0) \cos \alpha + p(y - y_0) \sin \alpha + t(z - 2p) + pt^3 = 0 \quad (34)$$

The single positive root of Eq. (34) is obtained by means of the Newton–Raphson iterative procedure. This requires a starting value for t, for which the vertical distance, z, on the infected root is used. Once t has been estimated, δ, the minimum distance, can be found by substituting the value for t into Eq. (33). If the minimum distance is less than the critical distance, cross-infection can occur. The relationships between inoculum density and the estimated number of cross-infections of *G. graminis* amongst seminal roots of five seedlings are shown for different values of the critical distance, in Fig. 5.

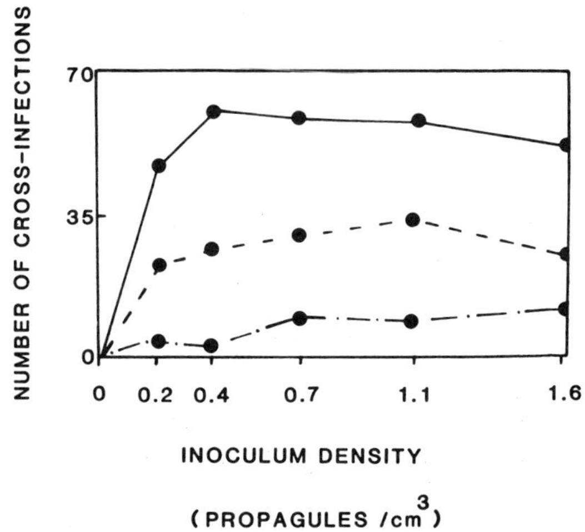

Fig. 5. Relationships between inoculum density and numbers of cross-infections on seminal roots of five seedlings of wheat when the minimum distance across which cross-infection can occur is 1.0 cm (—), 0.5 cm (- -), 0.2 cm (—·). The results were obtained by computer simulation.

IV. MODELS FOR HOST GROWTH

I have already noted that standard non-linear functions (Tinker, 1975) may be incorporated into models for root infection. Smith and Walker (1981) and Buwalda *et al.* (1982) used linear interpolation between successive observations of host growth in their models. Sanders and Sheikh (1983) used a polynomial function to incorporate a continuous function for root growth in a simulation model. Future work on modelling of subterranean epidemics must turn ever more frequently to models for host growth (Huisman, 1982). It is useful therefore to consider briefly the types of models that are available.

Most of the models published on the growth of host roots are concerned with change in root mass, and sometimes also root length over time. A few concern growth in space, usually depth. Greenwood *et al.* (1982) derive a simple empirical model to describe root density D with depth z. Suppose that

$$\mathrm{d}D/\mathrm{d}z = -qD$$

where q is a constant for any depth but varies with time. Integrating with respect to depth gives

$$D = D_0 \exp(-qz)$$

where D_0 is the root density at time $t = 0$ and depth $z = 0$. The total root length at time t between depths 0 and z is given by $L(z,t)$ where

$$L(z,t) = \int_0^z D \, \mathrm{d}z$$

so,

$$L(z,t) = D_0[1 - \exp(-qz)]/q$$

and when $z = \infty$, $L(z,t)$ is the total length of root, L_t:

$$L_t = D_0/q$$

An earlier form of the model to describe the proportion of root at any given density is given by Gerwitz and Page (1974), who successfully tested the model over a wide range of crop plants.

Page and Gerwitz (1974) also proposed models based on diffusion equations to describe the three-dimensional root systems of isolated plants, row crops and swards. The model for isolated plants such as lettuce is given below for illustration. It is based on Fick's second law of diffusion (see e.g. Rubinow, 1975);

$$\partial D/\partial t = F \, \partial^2 D/\partial z^2 \tag{35}$$

where D is the root density in soil ($\mathrm{cm \cdot cm^{-3}}$), F is a constant and z is distance. Certain boundary and other conditions are necessary to solve Eq. (35). Suppose that diffusion (i.e. growth) occurs from a point source into a semi-infinite medium, i.e. one that extends infinitely in all directions except upwards and that the diffusing substance (roots) is liberated continuously at a constant rate q and R is a radial distance over which growth takes place, such that $R = (x^2 + z^2)^{1/2}$. Given $D = 0$, $R > 0$, $t = 0$ (Page and Gerwitz, 1974):

$$D = q(2\pi FR) \, \mathrm{erfc}[R/2(Ft)^{1/2}]$$

where erfc[ ] is the error function complement. The model assumes that individual roots grow at random but the net spread of roots is from regions of high concentration to ones of lower concentration and that the rate of penetration into a zone of lower concentration is proportional to the concentration of roots in the zone of higher concentration (Page and Gerwitz, 1974). The authors acknowledge the artificiality of their assumption of constant rate of expansion of the root system. Nevertheless, agreement of the model with experimental observations was good.

The diffusion equations apply to mass of root tissue with no distinction between root members, i.e. main root axes, first, second and third order laterals. Empirical descriptions of the increase over time in numbers and length of root members of single main root axes of barley are given by Hackett and Rose (1972a,b). The models include parameters for the following components of root growth: the extension of growth of each type of root member; the rate of extension of the branched region along a root member; the density of branching and the time of onset of branching of each order or root member. Hackett and Rose (1972b) investigated some of the inter-relationships amongst the variables in the models. The models, which are presented as discrete expressions, are well suited to computer simulation. The dynamics of the geometry of root growth are considered further by Lungley (1973). Using computer simulation rather than single analytical solutions, Lungley (1973) included orientation of root members in a model for the growth of root systems of single plants. Variability in the onset of branching and in rates of growth and branching could therefore be included. The model also allowed estimation of the spatial pattern of root density within contiguous 'cells' in a hypothetical volume of soil. It does not estimate the number of contacts between roots but it is probable that some function relating root density to the probability of contacts could be incorporated. Such a model should prove highly valuable in mechanistic modelling of soil-borne pathogens. The models discussed so far describe the growth of roots in homogeneous media. Dexter (1978) and Hewitt and Dexter (1979) introduce probabilistic models for the growth of roots in heterogeneous media in which soil macrostructure is defined by the distribution of aggregates and voids. Hewitt and Dexter (1984) also describe the statistical properties of some of the root properties that influence the strength of penetration of aggregates.

An alternative approach to modelling of host growth that offers wide scope for the epidemiologist is in the mathematical description of the distribution of assimilates. Brouwer and De Wit (1969) describe a simple computer simulation model in which the pool of metabolites is apportioned amongst respiratory needs and vegetative growth of roots, shoots and leaves. The model includes simple negative feed-backs, approximating to logistic functions, but with variable rate parameters. A similar but more complex model for the growth and phenological development of a population of cotton plants was used by Guttierez *et al.* (1983) to model the effects of *Verticillium dahliae* in the crop. This simulation model is described briefly in section V.

Although computer simulation offers flexibility, it is not essential for modelling host growth and assimilate distribution. Greenwood *et al.* (1982) derive an analytical model for the root development of vegetable crops. The model is based on the assumption that the pool of photosynthates is partitioned between two components, one comprising leaves, stems and storage roots, and the other the fibrous roots. Within the fibrous roots, the photosynthates are used for growth and respiration but some is also lost due to leakage and microbial attack. The model is used to estimate the total length of fibrous root per unit area $L(t)$, at time t from emergence:

$$\ln L(t) = \gamma_j + \beta_j \ln W - \mu t \tag{36}$$

where W is total plant dry weight excluding fibrous roots, γ_j and β_j are parameters that depend on the crop (j) and μ is a parameter that describes the loss of organic carbon from roots by leakage and microbial attack. The parameters β and δ are complex (Greenwood *et al.*, 1982). β is itself a function of three parameters, one reflecting the environmental effects on root-to-shoot ratio, the others being constants of proportionality for the conversion of metabolites into root and non-root tissue, respectively; γ is an additive function of parameters for leakage, shoot weight and root length at time $t = 0$. The units of μ are grams per gram of root per unit time. Greenwood *et al.* (1982) obtained a value for μ that implied a constant loss of 3% of carbon in the roots per day in seven species of vegetable for which the model was tested. Inclusion of the component for loss from roots offers possibilities for further study of the effects of pathogens on crop growth. Fitting of the model to experimental data was by the method of least squares. The authors did, however, test several subjective forms of the model given in Eq. (36), for example by holding γ constant over crops and by omitting the term in t. While the full model gave the best fit, i.e. explained most of the variation in $\ln L(t)$, Greenwood *et al.* (1982) note that other models, including, for example, a term in $\ln(W^2)$ rather than one in time, describe the data equally well. This is a salutary reminder that correlation does not necessarily imply causation.

Much modelling of root systems has been directed at nutrient uptake (see e.g. Barley, 1970; Baldwin *et al.*, 1973). A further alternative, yet to be explored but not attempted here, is to turn these models around so that they describe the negative effect on growth by the 'uptake' of plant pathogens.

V. ASSEMBLING THE COMPONENTS OF THE INFECTION CHAIN

The components of the infection chain are assembled implicitly in the non-linear and linear growth equations: to assemble the components explicitly, i.e. to synthesise the epidemic from its components, requires computer simulation. Computer simulation is a flexible strategy that is not constrained by the need to use equations with algebraically tractable solutions. The recursive facility of 'DO' loops enables equations to be integrated numerically. Consider Eq. (1) in which $dL_i/dt = SL_i(nL_t - L_i)$. The equation can be written as a difference equation, $\Delta L_i = SL_i(nL_t - L_i)\Delta t$. If the initial length of root $L_{t(0)}$, and infected root, $L_{i(0)}$, are known, the amount of infection at time t may be estimated recursively, as follows:

$$
\begin{aligned}
t = 0{:}\;\; & L_{i(0)} = L_{i(0)} \\
t = 1{:}\;\; & L_{i(1)} = L_{i(0)} + \Delta L_i = L_{i(0)} + SL_{i(0)}(nL_{t(0)} - L_{i(0)})\Delta t \\
& \;\vdots \qquad \vdots \\
t = t_j{:}\;\; & L_{i(j)} = L_{i(j-1)} + SL_{i(j-1)}(nL_{i(j-1)} - L_{i(j-1)})\Delta t
\end{aligned}
$$

The variable L_t may be estimated by another recursive relationship. The parameters S and n may be changed deterministically according to environmental factors, such as temperature, or host factors, such as age of roots. Alternatively, or additionally, S, say, may be treated as a random variable. Values of S are then obtained from a random number generator for a given probability distribution with specified moments. This method is known as Monte Carlo simulation. It is a powerful but simple means of introducing stochastic variation into models (Sall, 1980), whereby the output of models for any given input is not pre-determined but may vary each time the model is run.

Yet more flexibility is given to computer simulators by the use of decision statements, such as the 'IF' statement. Suppose that infection of roots ceases when temperature is below a certain threshold (THRESH); an 'IF' statement can be used to stop infection:

```
IF (TEMP.LT.THRESH)DELTAL = 0.0
```

The very flexibility of computer simulation due to the accumulation of subroutines, stochastic variables and decision statements may also be a disadvantage. It can make the simulator 'unwieldy' and difficult to test (Zadoks, 1978). The philosophy of testing computer simulators is discussed by Teng (1981).

To date three computer simulators involving soil-borne, parasitic fungi have been published. The fungi and hosts are *Fusarium oxysporum* on Douglas fir (Bloomberg, 1979a); *G. caledonius*, a v-a-m fungus, on onions (Sanders and Sheikh, 1983); and *V. dahliae* on cotton (Guttierez *et al*., 1983). Rather than discuss each simulator in detail, I propose instead to indicate how the three components discussed in section III, primary infection, secondary infection and host growth, have been variously included in the models. The objective is to demonstrate the variety in approach possible with this type of modelling.

Bloomberg (1979a) simulated the infection of roots, by *F. oxysporum*, of up to 100 seedlings of Douglas fir for a period of 180 days. The simulator was updated at intervals of 5 days, according to temperature expressed as values for heat summation in degree hours per day. The principal output variables comprised the numbers of healthy, infected but living, and damped-off seedlings in each 5-day period. In addition the following were estimated: the total numbers and lengths of first- and second-order laterals, whether or not they were diseased and whether by primary or secondary infection.

Primary infection by *F. oxysporum* is restricted to the root tips (Bloomberg, 1979a; Nyvall and Hagland, 1972). Root growth was described in a three-dimensional lattice of contiguous cells and primary infection occurred when an infective propagule and root tip coincided in a cell. The occurrence of the propagules was random. The size of lattice cells reflected the density and stochastic variation in size of propagules. Bloomberg (1979a) used the observed frequency distribution of the propagules, noting that the probability of infection varied with size.

Secondary infection was described by lesion growth which in turn was estimated by regression of linear growth of *F. oxysporum* within roots on periodic heat sum. Lesions on second-order laterals invariably extended into parent first-order laterals. Importantly, cross-infection between seedlings is not thought to occur (Bloomberg, 1979a). This simplifies enormously what is nevertheless a complex and detailed model.

The growth of the host was modelled in detail. Increments in total root length were estimated from an empirical function of temperature, again expressed as heat summation. The growth of the taproot was estimated as a proportion of the total root length. Similarly, the numbers of first- and second-order laterals were estimated by regressions on the total length of the next higher order root members. The initial lengths of individual root members were estimated by random selection from a probability distribution. The location and orientation of root members were specified deterministically: for example, all second-order laterals were assumed to grow at right angles to the parent member. Radial symmetry in the horizontal plane was assumed. Thus Bloomberg (1979a) modelled the three-dimensional growth of the root systems of individual seedlings, albeit in a two-dimensional representation.

Bloomberg's model typifies the mechanistic type of simulator which models and records in detail the life cycle of the pathogen and growth of the host. It is important to note that it includes stochastic components for host growth and primary infection. Bloomberg (1979a) validated the model and then investigated the relative importance of some of the components in the progress of the disease (Bloomberg, 1979b). This was done by a simple form of sensitivity analysis in which the effects of varying input parameters by factors such as 0.5, 1.0, and 2.0 while holding other parameters constant were examined. The most important factors in simulated root infection were those affecting size and number of infection sites (root tips), size and density of propagules of inoculum and rate of lesion expansion. The results are probably not surprising. The value of this method of simulation lies not so much in these results but in the synthesis of disparate information on a host–pathogen system into a logical whole and the possibility this gives of screening many hypotheses prior to experimentation.

Sanders and Sheikh (1983) used a much less detailed simulator to model the infection of onion roots by *G. caledonius*. The model, which was deterministic, simulated infection within pots for a period of 50 days. The model was updated at intervals of 1 day to take account of host growth, fungal spread and the effects of infection on the growth of the host. The principal output variables were the daily change in the proportion of root length infected, the numbers of separate infections, phosphorus content, relative growth rate of roots and total dry weight of plants. The model is similar in certain respects to those of Walker and Smith (1984) and Buwalda *et al.* (1982).

The model does not distinguish between primary and secondary infections but stimulates changes in the total numbers of non-overlapping infections and in the length of infected root. In this respect, it resembles the model of Walker and Smith (1984). The potential number of new infections per day is estimated by a relative multiplication rate ($n^{-1}\ \mathrm{d}n/\mathrm{d}t$). The latter is obtained after differentiation of a polynomial regression of the number of non-overlapping infections (n) on time, obtained from experimental observations. The 'actual' number of non-overlapping infections is given after correction for the proportion of root length that remains infectible. Each infection is assumed to grow at a constant rate for a limited period, after which expansion stops.

The model for host growth is based upon the distribution of assimilates. It is similar in principal to that of Brouwer and De Wit (1969) but with phosphorus uptake as the rate-determining factor. Sanders and Sheikh (1983) assumed that a rate of uptake of phosphorus and hence the total phosphorus content of the plants varied in direct proportion to the proportion of roots that were mycorrhizal. Total dry weight of plants was estimated from the phosphorus content of plants by regression. Total dry weight of roots was estimated as a proportion of the total plant dry weight and total length of roots as a proportion of root dry weight. Daily increments of phosphorus content were estimated from the absolute amounts of infected and uninfected roots.

Sanders and Sheikh (1983) obtained good agreement with experimental data on the progress of infection and the growth response to mycorrhizal infection. The authors acknowledge, however, that they adjusted parameters and initial values for the spread of infection to give satisfactory agreement between predicted and measured estimates of the proportion of root length infected. The simulator is elegant in its comparative simplicity but it is arguably less of a synthesis of analytical hypotheses than that of Bloomberg (1979a). More comparative work on alternative methods of simulation modelling is needed.

The third simulation model is mathematically the most advanced of the three. Guttierez *et al.* (1983) coupled a sub-model for growth and development of a population of cotton plants (Gutteriez *et al.*, 1975; Wang *et al.*, 1977). The objective was to model the effect of disease (rather than infection) on the phenological development and yield of the crop. The model is deterministic with principal inputs, host density and inoculum density. Host growth accounts for most of the model. Primary and secondary infections are not modelled. Instead, the rate of increase in diseased plants is estimated from a regression function of inoculum density and a scalar variable, used to summarise the ability of a given pathotype of *V. dahliae* to cause disease. Physiological time in day–degrees is used. The effects of disease are included in models for the competitive allocation of metabolite for respiration, reproductive growth (fruits) and vegetative growth (roots, stems, leaves) within individual plants. Guittierez *et al.* (1983) note that the model for host growth is a continuous analogue of the Leslie Matrix Model which is widely used in ecological work for populations with age structure. The reduction in photosynthate production caused by disease is estimated on individual plants by a simple function that is directly related to the concomitant reduction of leaf mass over time and to inhibition of vascular transport. Compensation by healthy plants is included as a density-dependent process. The simulator involves the use of partial differential equations to model the components of host growth. Consideration of the population dynamics of the pathogen is minimal and elementary, yet the simulator works satisfactorily in modelling the effects of the pathogen on crop growth.

VI. CONCLUDING REMARKS

I have considered widely differing approaches to modelling in the foregoing discussion. In common with modelling of aerial pathogens, two broad strategies for modelling of soil-borne pathogens are apparent. They are distinguishable by the degree of detail aspired to in the models. Thus, the progress of disease may be summarised in relatively few parameters in growth curve analysis, or it may be dissected and modelled separately in components of the infection chain which are then reassembled by computer simulation. The choice of strategy must reflect the objectives of the investigation but will also be subject to constraints of mathematical expertise. Computer simulation is a very flexible strategy, with the powerful facility of introducing variation into models by the use of stochastic variables. It is logically suited to the synthesis of models for the components of the infection chains of specific host-pathogen systems. Computer simulation is disadvantaged by involving detailed models that are difficult to test and often difficult to read. Much more thought needs to be given to the methodology for writing computer simulation models and to the verification and validation of these models. The growth curve models have the advantage of wide applicability to a range of host–pathogen systems. Growth curves, too, may be flexible, especially with the introduction of more parameters, either in a single or in related equations. They are disadvantaged by lack of biological detail.

The infection chain is, I propose, a valuable concept to use in the construction of models for disease progress of soil-borne pathogens. Our knowledge about the infection chain, however, is uneven: much more is known about primary infection and indeed also about survival than is known about secondary infection. More detailed work on the mechanics of disease spread is needed.

Two phenomena, host growth and spatial pattern, have recurred in the consideration of models both for overall disease progress and for components of the infection chain. Host growth impinges in three important ways on the progress of disease of soil-borne pathogens. The rate of growth of the host limits the rate of primary infection. The geometry of root growth influences the rate of secondary infection and the interaction between infection and host growth influences the nutritional status of the host. This, in turn, influences not only yield but also the probability of further infections arising on the host. It is probable, therefore, that models for host growth will occupy an increasingly central position in modelling of disease progress of soil-borne pathogens.

The case for incorporation of the spatial pattern of inoculum into models for disease progress is less apparent but I suggest that it is still strong. The spatial pattern of inoculum influences the spatial pattern of primary infection. This in turn influences the pattern of secondary infection and the interaction with host growth. Randomness of occurrence of inoculum and infection is a much-invoked assumption in the construction of epidemiological models yet it is known to be at variance with both intuition and observation (see e.g. Gilligan, 1983, 1985b). It might be argued that the averaging effect of large samples justifies the assumption of randomness. We are dealing, however, with populations that are relatively immobile and so the effects of clustering of inoculum would seem to have a strong claim for inclusion in explanatory models for disease progress. Clustering of propagules may be treated deterministically in the same way as for randomness. Inclusion of the spatial pattern of inoculum in epidemiological models does not of itself, therefore, imply that the model is stochastic. Given, however, the natural variation evident in epidemics of soil-borne pathogens (see e.g. Campbell and Pennypacker, 1980; Campbell *et al.*, 1980a,b; Gilligan, 1982a; Taylor *et al.*, 1981) it is probable that stochastic models will play an increasingly prominant role in epidemiological modelling of these systems.

VII. APPENDIX

Glossary of symbols used to represent principal random variables and parameters in text.

A	Frequency of infections per unit length of root per unit time.
a	Searching efficiency of a parasitoid or predator.
a'	Instantaneous search rate (searching efficiency per unit time).
α	Angles of rotation of seminal roots of wheat about the crown.
α,β	1. Parameters of function for ψ_2 (see below).
	2. Proportions of total populations of prey and predator in a given volume.
B	Rate of spread of mycelium within a root from a single infection point.
b	Proportion of males to females of larvae within roots, per unit length of root.
β,γ	Parameters relating environmental factors and distribution of assimilates within plants to the expected length of fibrous roots.
C	Proportion of nematode eggs that fail to hatch.
c	Compensatory growth of a plant in response to nematode damage.
D	Root density (length per unit volume).
D_0	Root density at time $t = 0$ and depth $z = 0$.
δ	Distance between a point on an infected root and any point on an uninfected root.
F	Number of juvenile female nematodes per unit length of root.
F'	Number of juvenile female nematodes per gram of soil.
g	Relative growth rate of roots.
H	Proportion of hatched nematode larvae that invade roots.
h	Total length of roots exposed to nematodes
θ	Probability of occurrence of a propagule of inoculum within a pathozone.
k	Index of aggregation.
L	Length of a single, main root axis.
L_i	Length of infected root per unit volume of soil.
L_t	Total length of root per unit volume of soil.
$L_{t(min)}$	Yield (length of root) given a very large initial density of nematodes.
$L_{t(max)}$	Yield (length of root) in the absence of nematodes.
$L(t)$	Length of fibrous root per unit area at time t.
$L(z,t)$	Total length of root between depths 0 and z at time t.
l	Proportion of total length of root infected.
l_∞	Asymmptotic value of l.
m	Index of degree of interference amongst predators and parasites.
m'	Average number of eggs produced per female nematode.
m_∞	Apparent mean length of an infection unit when infection is asymptotic.
μ	Rate of loss of organic carbon from roots by leakage and microbial attack.
N	Number of hosts.
N_d	Number of diseased hosts.
N_e	Number of encounters between parasites and hosts.

N_p	Number of parasitised hosts.
N_u	Number of uninfected hosts.
n	1. Maximum proportion of root that can be infected. 2. Number of non-overlapping infections on roots.
P	1. Total number of propagules of inoculum. 2. Density of propagules of inoculum.
P_i	Initial density of propagules of inoculum.
P_f	Final density of propagules of inoculum.
P_r	Number of larvae within roots per unit length of roots.
P_t	Density of parasites at time t.
ϕ	Probability that a propagule of inoculum occurs in a pathozone and infects the host.
g	Decline in root density with depth.
R	1. Outer radius of pathozone. 2. Radial distance over which growth of roots of a single plant takes place.
s	Rate of increase of infection.
T	1. Threshold density of nematodes below which yield of plants is unaffected. 2. Time during which hosts are susceptible to attack by parasites and predators.
T_h	Time spent 'handling' hosts by parasites and predators.
T_s	Time available for 'searching' for hosts by parasites and predators.
U	Number of entry points (or infection units) per plant.
u	Number of infection units per unit length of root.
u_∞	Asymptotic value of u.
V	Volume of infested soil.
v	Volume of pathozone.
W	Total plant dry weight excluding fibrous roots.
x,y,z	Coordinates of point in soil.
Y	Level of disease.
ψ	Probability of propagule infecting host conditional upon its occurrence in the pathozone.
ψ_1	Probability that centre of a propagule occurs in annulus described by radii r and $r + \Delta r$.
ψ_2	Probability that infection can occur from a propagule located between r and $r + \Delta r$ from the central axis of an infection court.
z	1. Expected yield when nematode density exceeds a certain threshold T 2. Depth in soil.

REFERENCES

Ashton, W. D. (1972). "The Logit Transformation with Special Reference to its Uses in Bioassay." Griffin, London.

Baker, R. (1971). *Phytopathology* **61**, 1280–1292.

Baker, R. and Drury, R. E. (1981). *Phytopathology* **71**, 363–372.

Baker, R. and Wijetunga, C. (1981). *In* (D. T. Wicklow and G. C. Carrol, eds.), pp. 249–261. Dekker, New York.

Baldwin, J. P., Tinker, P. B. and Marriott, F. H. C. (1971). *J. Appl. Ecol.* **8**, 543–554.

Baldwin, J. P., Nye, P. H. and Tinker, P. B. (1973). *Plant Soil* **38**, 621–635.

Barley, K. P. (1970). *Adv. Agron.* **22**, 159–201.

Barley, K. P. and Sedgeley, R. H. (1961). *Aust. J. Sci.* **23**, 297–298.
Berger, R. D. (1981). *Phytopathology* **71**, 716–719.
Bloomberg, W. J. (1979a). *Phytopathology* **69**, 74–81.
Bloomberg, W. J. (1979b). *Phytopathology* **69**, 1072–1077.
Brouwer, R. and De Wit, C. T. (1969). "Root Growth." Butterworth, London.
Burdon, J. J. and Chivers, G. A. (1975). *Ann. Appl. Biol.* **81**, 135–143.
Buwalda, J. G., Ross, G. J. S., Stribley, D. P. and Tinker, P. B. (1982). *New Phytol.* **91**, 669–682.
Campbell, C. L. and Pennypacker, S. P. (1980). *Phytopathology* **70**, 521–525.
Campbell, C. L. and Powell, N. T. (1980). *Prot. Ecol.* **2**, 177–182.
Campbell, C. L., Madden, L. V. and Pennypacker, S. P. (1980a). *Phytopathology* **70**, 152–155.
Campbell, C. L., Pennypacker, S. P. and Madden, L. V. (1980b). *Phytopathology* **70**, 487–491.
Causton, D. R. (1977). "A Biologist's Mathematics." Arnold, London.
Cox, G. C. and Sanders, F. E. (1974). *New Phytol.* **73**, 901–912.
Crowe, F. J. and Hall, D. H. (1980). *Phytopathology* **70**, 70–73.
Dexter, A. R. (1978). *J. Soil. Sci.* **29**, 102–116.
Drury, R. E., Baker, R. and Griffin, G. J. (1983). *Phytopathology* **73**, 1351–1354.
Ferriss, R. S. (1981). *Phytopathology* **71**, 1229–1231.
Ferriss, R. S. (1982). *Phytopathology* **72**, 1397–1403.
Ferriss, R. S. (1983). *Phytopathology* **73**, 1355–1357.
Fitzpatrick, E. G. and Rose, L. E. (1936). *Am. Soil Surv. Assoc. Bull.* **17**, 136–145.
Fulton, W. C. (1979). *Phytopathology* **69**, 1162–1164.
Fusseder, A. (1983). *Plant Soil* **73**, 441–445.
Gäumann, E. (1951). "Pflanzliche Infectionslehre," 2nd Ed. Birkhäuser, Basel.
Gerwitz, A. and Page, E. R. (1974). *J. Appl. Ecol.* **11**, 773–782.
Gilligan, C. A. (1979). *Phytopathology* **69**, 782–784.
Gilligan, C. A. (1980a). *Soil Biol. Biochem.* **12**, 513–514.
Gilligan, C. A. (1980b). *Trans. Br. Mycol. Soc.* **75**, 419–424.
Gilligan, C. A. (1982a). *J. Agric. Sci.* **99**, 461–464.
Gilligan, C. A. (1982b). *Trans. Br. Mycol. Soc.* **73**, 193–200.
Gilligan, C. A. (1983). *Annu. Rev. Phytopathol.* **21**, 45–64.
Gilligan, C. A. (1985a). *Phytopathology* **75**, 61–67.
Gilligan, C. A. (1985b). *In* "Populations of Plant Pathogens: Their Dynamics and Genetics" (M. S. Wolfe and C. E. Caten, eds.). Blackwell, Oxford, in press.
Greenwood, D. J., Gerwitz, A., Stone, D. A. and Barnes, A. (1982). *Plant Soil* **68**, 75–96.
Gregory, L. V., Ayers, J. E. and Nelson, R. (1981). *Phytopathol. Z.* **100**, 135–142.
Guttierez, A. P., Falcon, L. A., Loew, W. and Leipzig, P. A. (1975). *Environ. Entomol.* **4**, 125–136.
Guttierez, A. P., DeVay, J. E., Pullman, G. S. and Frieberthauser, G. E. (1983). *Phytopathology* **73**, 89–95.
Gyani, B. P. (1945). *J. Phys. Chem.* **49**, 442–453.
Hackett, C. and Rose, D. A. (1972a). *Aust. J. Biol. Sci.* **25**, 669–679.
Hackett, C. and Rose, D. A. (1972b). *Aust. J. Biol. Sci.* **25**, 681–690.
Hassell, M. P. (1976). "The Dynamics of Competition and Predation." Arnold, London.
Hassell, M. P. (1978). "The Dynamics of Arthropod Predator Prey Systems." Princeton Univ. Press, Princeton, New Jersey.
Hassell, M. P. and May, R. M. (1973). *J. Anim. Ecol.* **42**, 693–726.
Hassell, M. P., Lawton, J. H. and Beddington, J. R. (1977). *J. Anim. Ecol.* **46**, 249–262.
Hau, F. C., Campbell, C. L. and Beute, M. K. (1982). *Plant Dis.* **66**, 568–571.
Henis , Y. and Ben-Yephet, Y. (1970). *Phytopathology* **60**, 1351–1356.
Hewitt, J. S. and Dexter, A. R. (1979). *Plant Soil* **52**, 325–343.
Hewitt, J. S. and Dexter, A. R. (1984). *Plant Soil* **77**, 39–51.

Holling, C. S. (1959). *Can Entomol.* **91**, 293–320.
Huisman, O. J. (1982). *Annu. Rev. Phytopathol.* **20**, 303–327.
Jones, F. G. W. (1957). *Nematologica* **2**, 257–272.
Jones, F. G. W. and Kempton, R. A. (1978). *In* "Plant Nematology." MAFF AS Publ. GD/1. HM Stationary Office, London.
Jones, F. G. W. and Perry, J. N. (1978). *J. Appl. Ecol.* **15**, 349–371.
Jones, F. G. W., Parrott, D. M., and Ross, G. J. S. (1967). *Ann. Appl. Biol.* **60**, 151–171.
Kendall, M. G. (1976), "Time Series," 2nd Ed. Griffin, London.
Kranz, J. (1974). *In* "Epidemics of Plant Diseases: Mathematical Analysis and Modelling" (J. Kranz, ed.), pp. 1–6. Springer-Verlag, Berlin and New York.
Kushalappa, A. C. and Ludwig, A. (1982). *Phytopathology* **72**, 1373–1377.
Last, F. T., Ebben, M. H., Hoare, R. C., Turner, E. A. and Carter, A. R. (1969). *Ann. Appl. Biol.* **64**, 449–459.
Lownsbery, B. F. and Peters, B. G. (1955). *Phytopathology* **45**, 163–167.
Lungley, D. R. (1973). *Plant Soil* **38**, 145–159.
Madden, L. V. (1980). *Prot. Ecol.* **2**, 159–176.
May, R. M. (1978). "Stability and Complexity in Model Ecosystems." Princeton Univ. Press, Princeton, New Jersey.
Melhuish, F. M. (1968). *Ann. Bot. (London)* **32**, 15–22.
Morrall, R. A. A. and Verma, P. R. (1981). *Can. J. Plant. Pathol.* **3**, 182–183.
Neter, J. and Wasserman, W. (1976). "Applied Linear Statistical Models: Regression Analysis, Analysis of Variance and Experimental Designs." Irwin, Homewood, Illinois.
Nicholson, A. J. (1933). *J. Anim. Ecol.* **2**, 132–178.
Nicholson, A. J. and Bailey, V. A. (1935). *Proc. Zool. Soc. London* **1935**, 551–598.
Nyvall, R. F. and Haglund, (1972). *Phytopathology* **62**, 1419–1424.
Page, E. R. and Gerwitz, A. (1974). *Plant Soil* **41**, 243–254.
Pennypacker, S. P., Knoble, H. D., Antle, C. E., and Madden, L. V. (1980). *Phytopathology* **70**, 232–235.
Perry, J. N. (1978). *J. Appl. Ecol.* **15**, 781–787.
Peters, B. G. (1961). *J. Helminthol.*, R. T. Leiper 80th birthday vol., pp. 141–150.
Pfender, W. F. (1982). *Phytopathology* **72**, 31–32.
Pfender, W. F. and Hagedorn, D. J. (1983). *Phytopathology* **73**, 1109–1113.
Pielou, E. C. (1977). "Mathematical Ecology." Wiley, New York.
Plaut, J. L. and Berger, R. D. (1981). *Phytopathology* **71**, 917–921.
Pullman, G. S. and DeVay, J. E. (1982). *Phytopathology* **72**, 549–554.
Punja, Z. K. and Grogan, R. G. (1981). *Phytopathology* **71**, 1099–1103.
Reynolds, K. M. and Bloomberg, W. J. (1982). *Can. J. For. Res.* **12**, 493–498.
Robinson, R. A. (1969). *Rev. Appl. Mycol.* **48**, 593–606.
Rogers, D. J. (1972). *J. Anim. Ecol.* **41**, 369–383.
Rouse, D. I., MacKenzie, D. R. and Nelson, R. R. (1981). *Phytopathol. Z.* **100**, 143–149.
Rubinow, (1975). "Introduction to Mathematical Biology." Wiley, New York.
Sall, M. A. (1980). *Pflanzenkr. Pflanzenschutz* **87**, 397–403.
Sanders, F. E. and Sheikh, N. A. (1983). *Plant Soil* **71**, 223–246.
Seinhorst, J. W. (1965). *Nematologica* **10**, 87–94.
Seinhorst, J. W. (1970). *Annu. Rev. Phytopathol.* **8**, 131–156.
Smith, S. E. and Walker, N. A. (1981). *New Phytol.* **89**, 225–240.
Solomon, M. E. (1949). *J. Anim. Ecol.* **18**, 1–35.
Stack, R. W. (1980). *Can J. Plant. Pathol.* **2**, 187–193.
Stanghellini, M. E., von Bretzel, P., Kronland, W. C. and Jenkins, A. D. (1982). *Phytopathology* **72**, 1481–1485.

Stanghellini, M. E., Stowell, C. E., Kronland, W. C. and von Bretzel, P. (1983). *Phytopathology* **73**, 1463–1466.

Taylor, J. D., Griffin, G. D. and Garren, K. H. (1981). *Phytopathology* **71**, 1297–1302.

Teng, P. S. (1981). *Pflanzenkr. Pflanzenschutz* **88**, 49–63.

Tinker, P. B. (1975). *Symp. Soc. Exp. Biol.* **29**, 325–349.

Trudgill, D. L. (1967). *Nematologica* **13**, 263–272.

Turner, M. E., Blumstein, B. A. and Sebaugh, J. L. (1969). *Biometrics* **25**, 577–580.

Turner, M. E., Bradley, E. L., Kirk, K. A. and Pruitt, K. M. (1976). *Math. Biosci.* **29**, 367–373.

Van der Plank, J. E. (1975). "Plant Diseases: Epidemics and Control." Academic Press, New York.

Walker, N. A. and Smith, S. E. (1984). *New Phytol.* **96**, 55–69.

Wang, Y., Guttierez, A. P., Oster, G. and Daxl, R. (1977). *Can. Entomol.* **109**, 1356–1374.

Zadoks, J. C. (1978). *In*: "Plant Disease: An Advanced Treatise" (J. G. Horsfall and E. B. Cowling, eds.), Vol. 2, pp. 63–96. Academic Press, New York.

5

CONSTRUCTION OF DISPERSAL MODELS

H. A. McCartney and B. D. L. Fitt

Rothamsted Experimental Station, Harpenden, Hertfordshire, England

I. INTRODUCTION

In models of pathogen life cycles dispersal is frequently the weak link (Waggoner, 1983). Unlike infection or sporulation, dispersal, by definition, involves more than one location. Consequently, pathogen dispersal is difficult to measure under field conditions or to simulate in controlled experiments. The resultant lack of experimental information makes the construction of dispersal models difficult.

We shall confine our discussion to models of dispersal of plant pathogen propagules by wind and rain, the two main natural agents in spread of plant diseases. We shall not consider models of pathogen spread by insects, flowing water, soil or man. Much of the chapter will be concerned with the aerial dispersal of fungal spores, as fungi are the most common and most studied causes of plant diseases. However, the same principles apply to dispersal of some bacterial and viral plant pathogens when they are dispersed on air-borne debris or in spash droplets.

A. Importance of Spore Dispersal

Early in an epidemic, disease is often seen as discrete patches (foci) in crops. As an epidemic develops, the total amount of disease in a crop increases in two ways: the foci intensify as further infections occur on plants which are already infected (increased disease severity) and the foci spread as infections occur on previously healthy plants (increased disease incidence). The rate at which disease foci intensify and spread depends on many environmental and biological factors, one of which is how far viable infective spores travel from infected plants. If spores are carried long distances (tens of metres), disease foci should expand rapidly but intensify slowly; if spores are carried only short distances (a few centimetres), disease foci should expand slowly, but intensify rapidly. Spore dispersal may be important in the initial infection of fields and the spread of disease over large regions; in such cases the distances involved range from a few metres to hundreds of kilometres. Indeed, spore dispersal can probably be ignored only when disease is uniformly spread throughout a crop and only average amounts of disease in the crop are of interest.

B. Spore Dispersal Processes

Spore dispersal can be divided into three phases: release or removal, transport and deposition. Some spores are released by an active liberation mechanism, such as 'flick-off', in which spores are thrown off from the host by rapid changes in sporophore shape in response to changes in relative humidity (Ingold, 1971). Others are released by puff or tap mechanisms initiated by raindrops, i.e. by the puff of air which precedes a drop spreading on a host surface or by the shaking of the host caused by the impact of the drop (Hirst and Stedman, 1963). Such spores are released at all wind speeds. Others are blown from the host by wind or shaken off by the action of wind on foliage, and as their removal may require considerable force, it may occur only in gusts (Aylor *et al.*, 1981). Spores may be removed directly from their host by rain-splash or drip-splash droplets. Small, spore-carrying splash droplets may be carried considerable distances by the wind in turbulent eddies, but larger splash droplets follow ballistic trajectories little affected by turbulence and travel shorter distances (Fitt and Bainbridge, 1983). Spore deposition is by impaction or sedimentation. Splash-dispersed spores are usually in a water film when they reach a new host and are likely to be retained after deposition; dry-dispersed spores are more likely to bounce off.

C. Dispersal Models

Dispersal models may be either empirical or physical. Empirical models attempt to fit mathematical formulae to experimentally measured dispersal gradients. Physical models try to describe dispersal mathematically, using theories based on the laws of physics. Physical models have been widely used to describe the dispersal of air pollutants and the movement of water vapour and CO_2 into and away from crops. We shall show how these empirical and physical models are constructed and what limitations must be placed on their use.

II. EMPIRICAL MODELS

Empirical models of dispersal gradients are descriptive and not interpretative (Gregory, 1968). They may describe a gradient more or less accurately, but cannot explain how it arose. The empirical models we shall discuss are generally concerned with short-distance dispersal (up to 1 km), since only rarely can gradients be measured over longer distances.

A. Types of Model for Dispersal Gradients

The most commonly used empirical models are the power law model and the simple exponential model. The power law model (Gregory, 1968) assumes that the number of spores deposited or number of infections produced (y) is inversely proportional to some power of the distance from the source (x):

$$y = ax^{-b} \tag{1}$$

where a and b are constants. The simple exponential model (Kiyosawa and Shiyomi, 1972) assumes that y decreases exponentially with distance:

$$y = c\exp(-dx) \tag{2}$$

where c is the number of spores deposited or amount of disease at the source and d is a constant. The coefficients b and d both relate to the rate of decrease in y with distance (gradient) but they differ between the two models; b is dimensionless in the power law model but d has the dimension of length^{-1} in the exponential model.

Strictly speaking, both Eqs. (1) and (2) are examples of the general exponential model (Lambert *et al.*, 1980)

$$y = \alpha\exp(-\beta x^n) \tag{3}$$

for the cases $n \to 0$ and $n = 1$, respectively.

The case with $n = 2$ is equivalent to the model of Kampmeijer and Zadoks (1977):

$$y = a/(2\pi s^2)\exp(-x^2/2s^2)$$

which assumes that there is a normal (Gaussian) distribution of spores round a point source, with a variance of s^2. The general exponential model [Eq. (3)] may give a better fit to a given set of data because the curve can take variable shapes (Lambert *et al.*, 1980) but it is doubtful whether such precision is needed in many cases. Therefore we shall confine our discussion to the more widely used power law and simple exponential models.

The distance x is measured in a direction at right angles to a line source (e.g. infected hedge) or the edge of an area source (e.g. an infected field). With a point source (e.g. an individual plant or the focus of an infection) the pattern of deposition is greatly influenced by wind direction and turbulence (McCartney and Bainbridge, 1984) and the value of the coefficient b or d depends upon the direction in which x is measured. The influence of wind direction can be removed if y is defined as the spore deposit or amount of disease integrated around the source at a distance x (Chamberlain, 1975; McCartney and Bainbridge, 1984):

$$y(x) = \int_0^{2\pi} d(x, \theta) d\theta$$

where $d(x, \theta)$ is the deposit per unit area at a distance x from the source and angle θ (measured from the mean wind direction).

It is easier to calculate the gradients b and d if logarithms of Eqs. (1) and (2) are used. For example the power law equation becomes

$$\ln(y) = \ln(a) - b \cdot \ln(x)$$

and the exponential equation becomes

$$\ln(y) = \ln(c) - dx$$

The coefficients b and d are then the slopes of the linearised equations. The parameters a and c depend on the source strength and are of interest only if two gradients within one experiment are being compared. The values of b and d are affected by the shape of the source and by the way in which y is measured (Gregory, 1968). The same empirical models can be fitted to disease expressed as the number of leaves infected, area of diseased foliage or fraction of plants infected; however, allowance may have to be made for the effects of multiple infections before analysis (Gregory, 1948; Jeger, 1983).

B. Comparison of Power Law and Exponential Models

There have been few comparisons between these two models in which both sets of parameters have been published. The work of Kable *et al.* (1980) on peach powdery mildew is a notable exception. Kiyosawa and Shiyomi (1972) use several sets of data given in Gregory (1945); the equivalent power law model parameters and coefficients of determination (r^2) can be calculated (Table I). Both models generally fitted these data reasonably well. Of 18 gradients, the power law model fitted the data better in 10 cases and the exponential model in 8. Ignoring differences in r^2 of less than 5%, these figures reduce to 6 and 4, respectively, the greatest difference being with *Pseudocercosporella herpotrichoides* (Oort, 1936) (Fig. 1).

McCartney and Bainbridge (1984) found that data for deposition of 20-μm droplets (simulating *Erysiphe graminis* spores) in barley crops were fitted equally well by the power law and exponential models. However, gradients of *E. graminis* spore deposition onto vertical cylinders within wheat (Fried *et al.*, 1979) and *Cronartium quercuum* spore concentration over open ground (Schmidt *et al.*, 1982) were represented better by the power law model than by the exponential model. This was also the case for wheat mildew disease (number of lesions) gradients (Fried *et al.*, 1979). On the other hand, disease (number of lesions) gradients of fusiform rust in oak seedlings were represented better by the exponential model (Schmidt *et al.*, 1982), perhaps because there was an elevated volume source, with variable wind direction. Disease incidence gradients of *Septoria nodorum* in spring wheat were fitted better by a power law model, whereas disease severity gradients were fitted better by an exponential model (Jeger *et al.*, 1983). These authors suggested that this was because outward spread of disease (incidence) was dependent on distance from the source whereas intensification of disease (severity) was not.

One advantage of the power law model is that the exponent b is independent of the units in which distance is measured (Gregory, 1968), unlike d of the exponential model. However, as x tends to zero the power law equation predicts values of y which tend to infinity and thus it generally overestimates deposition near to the source (Lambert *et al.*, 1980; McCartney and Bainbridge, 1984; Kable *et al.*, 1980), and so may be unsuitable for diseases with steep gradients. The exponential model tends to underestimate deposition near the source (Lambert *et al.*, 1980; McCartney and Bainbridge, 1984; Kable *et al.*, 1980; Schmidt *et al.*, 1982) but predicts finite values of y at all distances and is more easily incorporated into disease development models. Additionally, the equation has the property that the value of y falls to half in a fixed distance, α (the half-distance), no matter from where it is measured. The half-distance is related to the exponent d by $\alpha = 0.693/d$ and is an easily visualised measure of deposition gradient (i.e. the distance in which deposition or disease falls by half). Neither model should be extrapolated to distances beyond the observed gradient (Gregory, 1982). Few other generalisations can be made without more data on comparisons between the two models.

Table I. Use of the power law and exponential models to describe dispersal gradients in selected host pathogen systems.

Pathogen	Host	Power Law Model[a]			Exponential Model[a]			Reference
		ln(a)	b	r^2 (%)	ln(c)	d_1 (m)	r^2 (%)	
Cronartium ribicola	Pine	4.81	1.444	89.2	4.07	0.231[b]	96.1	Buchanan and Kimmey (1938)
Puccinia graminis	Couch grass	10.87	2.035	81.4	6 .49	0.078[b]	78.4	Johnson and Dickson (1919)
Oidium sp.	Peach							Kable *et al. (1980)*
	Orchard 1, 1974	9.39	1.923	97.9	3.67	0.039	94.5	
	Orchard 2, 1976	13.54	3.401	83.9	4.90	0.084	81.9	
	Orchard 2, 1976	11.58	3.000	95.9	4.06	0.079	94.4	
Puccinia graminis	Wheat	11.97	3.327	97.5	12.66	1.050[b]	98.4	Lambert (1929)
Phytophthora infestans	Potato							
	Focus 1	2.97	1.872	66.8	5.78	2.272[b]	92.8	Limasset (1939)
	Focus 2	4.01	1.490	74.4	5.99	1.489[b]	94.7	
Peronospora destructor	Onion	16.42	2.419	91.3	6.71	0.008[b]	94.9	Newhall (1938)
Pseudocercosporella herpotrichoides	Wheat	5.20	0.934	98.3	4.38	0.079[b]	59.4	Oort (1936)
Ustilago tritici	Wheat							
	North–east	5.04	0.691	85.3	4.65	0.083[b]	65.6	Oort (1940)
	South–east	3.64	0.649	56.3	3.00	0.034[b]	65.2	
	Location 2	6.56	0.768	95.9	6.94	0.284[b]	86.6	
Erysiphe graminis	Barley							
	Cv. 1	5.50	0.963	93.4	5.20	0.223[b]	84.5	Pape and Rademacher (1934)
	Cv. 2	6.06	0.640	84.2	5.87	0.152[b]	80.6	
	Location 2	10.86	1.436	53.5	4.96	0.008[b]	62.8	
Botrytis tulipae	Tulip							
	Bed 1	−0.25	1.408	97.1	3.04	4.696[b]	99.6	Wallace (1934)
	Bed 2	2.10	0.739	97.5	3.51	1.512[b]	81.7	

[a] Distance in metres.

[b] To obtain Kiyosawa and Shiyomi values multiply by 2.303 [i.e. ln(10)].

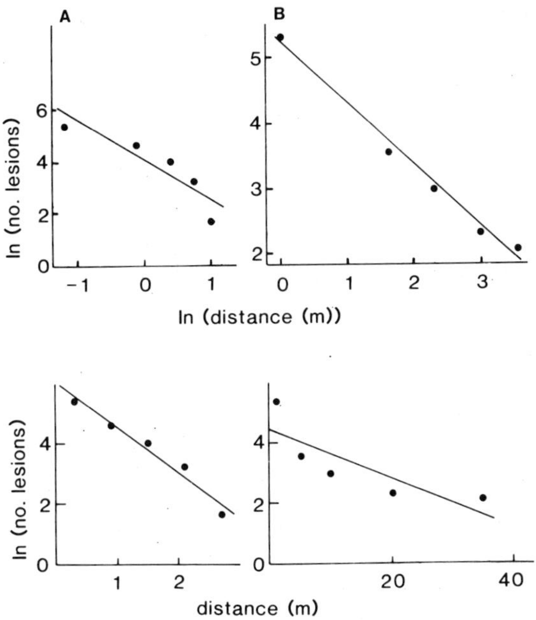

Fig. 1. Power law and exponential models of disease gradients: (A) potato blight (Limasset, 1939, focus 2); (B) eyespot (Oort, 1936).

C. Interpretation and Misinterpretation of Dispersal Gradients

The principles for interpreting dispersal gradients have been summarised by Gregory (1968, 1973, 1982); nevertheless, gradients are frequently misinterpreted. The observation of a spore dispersal gradient implies the existence of a local source, while the presence of a disease gradient implies the existence of a spore dispersal gradient, but not vice versa. The size, shape and height of the source all affect the gradient; for example gradients from point sources are generally steeper than those from line or area sources (Gregory, 1968) and ground level sources often produce steeper gradients than above-ground sources (Stedman, 1980b). Deposition from above-ground sources may increase with distance when there is a 'skip-distance' [a distance before deposition starts (Gregory, 1973)]. However, inoculum from a large number of distance sources will not produce a gradient (uniform deposition). A primary gradient results from one source releasing spores for a short time; secondary spread resulting from further generations of the pathogen (Cammack, 1958) will flatten a primary gradient. Gradients are influenced by local weather; for example, gradients of wind-borne pathogens are generally shallower when measured along the prevailing wind direction than in other directions.

Thus disease gradients with small values of *b* or *d* suggest proximity to an area source, secondary infection or a large background infection. Disease gradients with large *b* or *d* suggest that spore deposition gradients are also steep or that there may be no infection sites available (e.g. no susceptable hosts). However, it is difficult to draw conclusions from small differences in the values of *b* or *d* between pathogens, or between experiments with the same pathogen, because these gradients are affected by many different factors. That gradients are generally steeper within a crop than above open ground has been shown for spores of *Lycopodium clavatum* and *E. graminis* (Bainbridge and Stedman, 1979). Gradients of splash-dispersed spores are often steeper than those of wind-dispersed spores, both over open ground (Stedman, 1979) and within a crop (Stedman, 1980a). This difference may occur because most splash-dispersed spores are carried in ballistic splash droplets with diameters between 100 and 1000 μm which fall to earth in quite short distances (Fitt and Bainbridge, 1983), whereas wind-dispersed spores are usually smaller and are carried over longer distances. However, gradients may not differ greatly between wind-dispersed spores of different sizes, as in experiments comparing dispersal of *Lycopodium* (32 μm diameter) and *Bovista* (4 μm) spores (Stepanov 1935), and of *Lycopodium* and *Podaxis* (12 μm) spores (Sreeramulu and Ramalingam, 1961). Despite the number of investigations of spore deposition and disease gradients, there is little quantitative information about the effect of crop structure, spore type or wind on gradient formation.

A steep spore deposition or disease gradient does not imply that a source is ineffective at long distances (Gregory, 1982). The gradients result from the fraction of the spore cloud which is deposited and give no information about the fraction which remains air-borne. It is a paradox that meteorological factors, such as frictional turbulence and thermal convection, which cause steep gradients can lead to larger 'escape fractions' (the part of a spore cloud which reaches a height where it can disperse over long distances). Both Stepanov (1935) and Sreeramulu and Ramalingam (1961) estimated source strength and deduced that the fraction of spores which remained air-borne was greater for the smaller spores (*Bovista* and *Podaxis*) than the larger (*Lycopodium*), despite the similar deposition gradients. Another fallacy is to deduce from a change from a continuous disease gradient near to a source to an intermittent gradient further away that there are two different methods of spread; there may be or there may not (Gregory, 1982). A third error is to assume that a large difference between the amount of disease in an unsprayed control plot and a neighbouring sprayed plot in a field experiment implies that the treatment given to one plot does not affect its neighbours (Jenkyn *et al.*, 1979).

III. PHYSICAL MODELS

Spores, pollen and small water drops ($<100\ \mu m$) are carried in wind by turbulent eddies. Small particles are also subject to random molecular motions (Brownian diffusion) but these motions are small by comparison with turbulent eddies and may be neglected in estimating particle dispersal (Csanady, 1973). The phenomena of atmospheric turbulence and turbulent dispersal of gaseous and particulate pollutants have been studied experimentally and theoretically for many years (Pasquill, 1962, 1974; Tennekes and Lumley, 1972; Csanady, 1973; Harris, 1979; Nieuwstadt and van Dop, 1982). However, the results were seldom applied to plant pathogen dispersal until the late 1970s (Aylor, 1978, 1982; Legg and Powell, 1979; Itier and Pauvert, 1979; Aylor and Taylor, 1982; Aylor *et al.*, 1982; Legg, 1983a).

A physical model describing the dispersal of spores must include the effects of wind, turbulence and deposition. Thus any model requires appropriate information on wind, turbulence and deposition. Three atmospheric dispersal models have been applied to spore dispersal; gradient transfer theory (*K* theory), statistical (or Gaussian plume) and random walk models. The models vary in mathematical complexity and in the degree of sophistication in treating meteorological and biological variables. All three models can, with appropriate modifications, be applied to dispersal within and above crops over a wide range of distances. However, gradient transfer theory and Gaussian plume models are most applicable to spore transport above crops. In this section we will discuss the mathematical basis of each model and the meteorological information required for its application to spore dispersal. The mathematical description of deposition processes will also be considered.

A. Gradient Transfer Theory (*K* Theory)

1. The Diffusion Equation

Molecular diffusion of gases and Brownian diffusion of small particles in still air both obey Fick's Law (Csanady, 1973), i.e. the rate of diffusion depends on the concentration gradient of the diffusing material. The gradient transfer theory (*K* theory) of turbulent dispersal is based on an analogy with molecular diffusion (Taylor, 1915) and dispersal is described by the classical diffusion equation. For a cloud of particles the variation in concentration (C) in time and space is described in Cartesian co-ordinates by:

$$\frac{\mathrm{d}C}{\mathrm{d}t} = S(x,y,z) + \frac{\partial}{\partial x}\left(K_x \frac{\partial C}{\partial x}\right) + \frac{\partial}{\partial y}\left(K_y \frac{\partial C}{\partial y}\right) + \frac{\partial}{\partial z}\left(K_z \frac{\partial C}{\partial z}\right) + v_s \frac{\partial C}{\partial z} - u\frac{\partial C}{\partial x} - v\frac{\partial C}{\partial y} - w\frac{\partial C}{\partial z} \quad (4)$$

where u, v and w are the x, y and z components of mean wind speed, respectively, and K_x, K_y and K_z are the eddy diffusivities in the x, y and z directions, S is a source or sink term and v_s is the fall speed of the particles. The equation states that the rate of increase in concentration of particles in a small volume $(dx \cdot dy \cdot dz)$ equals the rate of production or loss of particles, S (for particle loss, S is negative) plus the rate of diffusion of particles into the volume $[\partial/\partial x(K_x \, \partial C/\partial x)$, etc.] plus the rate of sedimentation of particles into the volume $(v_s \, \partial C/\partial z)$ minus the rate of advection of particles out of the volume $(u \, \partial C/\partial x$, etc.).

When applied to turbulent diffusion Eq. (4) presents some difficulties. The rate of diffusion of a cloud of material depends on the relative sizes of the cloud and of the turbulent eddies; eddies smaller than the cloud cause it to grow slowly, eddies larger than the cloud transport it as a whole, whilst eddies of a similar size to the cloud disperse it most effectively. Thus as the cloud expands the larger eddies become more important and, as they contain more energy, the rate of diffusion increases (Csanady, 1973; Pasquill, 1974). Eddy diffusion coefficients, therefore, vary in time and space and are properties of the flow and not of the medium. To avoid the use of time-dependent diffusivities, Eq. (4) is strictly applicable only when the size of the cloud is greater than the characteristic size of the diffusing eddies. Thus it becomes invalid close to elevated point or line sources; it is, however, valid when applied to vertical diffusion from ground level sources (Pasquill, 1974; Hanna, 1982) or diffusion from volume sources such as a group of infected plants within a crop.

The full three-dimensional form of the diffusion–advection equation is extremely difficult to solve, but can be simplified by choosing the x axis to be the mean wind direction (Fig. 3); then diffusion in the x direction $[\partial/\partial x(K_x \, \partial C/\partial x)]$ can be neglected as particle transport by advection is much greater than by diffusion (Csanady, 1973; Pasquill, 1974). Except near field boundaries or during intense thermal convection, we can also assume that the vertical and cross-wind components of mean wind speed are zero ($v = w = 0$). Equation (4) then reduces to

$$\frac{dC}{dt} = S(x,y,z) + \frac{\partial}{\partial y}\left(K_y \frac{\partial C}{\partial y}\right) + \frac{\partial}{\partial z}\left(K_z \frac{\partial C}{\partial z}\right) + v_s \frac{\partial C}{\partial z} - u \frac{\partial C}{\partial x} \qquad (5)$$

If diffusion is from an infinite line source, or semi-infinite area source, orientated at right angles to the mean wind direction, cross-wind diffusion can be ignored and Eq. (5) simplifies to

$$\frac{dC}{dt} = S(x,z) + \frac{\partial}{\partial z}\left(K_z \frac{\partial C}{\partial z}\right) + v_s \frac{\partial C}{\partial z} - u \frac{\partial C}{\partial x} \qquad (6)$$

Additionally, if the source is emitting particles continuously and only time-average concentrations are of interest, then the time derivative may be ignored (steady state) and Eq. (6) becomes

$$u\frac{\partial C}{\partial x} = S(x,z) + \frac{\partial}{\partial z}\left(K_z\frac{\partial C}{\partial z}\right) + v_s\frac{\partial C}{\partial z} \tag{7}$$

Equation (7) is often adequate for describing the dispersal of plant pathogens.

All the above equations can be applied to spore dispersal both above crops and within crops. The solution of the chosen equation depends on the forms of K, u and S, which must be defined for all points in space (and time for time-dependent equations), and on the boundary conditions for the problem being solved. These variables and boundary conditions differ, in particular, between dispersal above and dispersal within crops.

2. Choice of Boundary Conditions

Generally we need only specify boundary conditions at large distances from the source ($x \to \infty$, $y \to \infty$), at the ground ($z = 0$) and at an upper boundary in the atmosphere. Particle concentration, C, is usually taken as tending to zero at large distances from the source. It is often assumed that particle concentration tends to zero at high altitudes ($C \to 0$ as $z \to \infty$). In the real atmosphere diffusion is severely limited at the top of the planetary boundary layer (PBL) where air flow is little affected by the ground surface (Fig. 2) (Fisher, 1979; Hanna, 1982). Thus, if particles do not cross the top of the planetary boundary layer, the boundary condition can be written as

$$K_z\frac{\partial C}{\partial z} + v_s C = 0 \text{ at } z = a \tag{8}$$

i.e. the total flux of particles is zero at height a, the top of the PBL (Maul, 1977; Scriven and Fisher, 1975; Nieuwstadt, 1980; Hanna, 1982). There are, however, several mechanisms which can transport particles through the top of the PBL (Fisher, 1979). This loss may be approximated, not wholly satisfactorily, by the use of a leakage velocity v_i (Maul, 1975; Llewellyn, 1983):

$$K_z\frac{\partial C}{\partial z} + v_s C = -v_i C \text{ at } z = a \tag{9}$$

v_i can be considered the effective upward velocity of particles leaving the boundary layer. In practice few spores dispersing from ground level reach the top of the boundary layer and Eq. (8) is usually an acceptable boundary condition.

When the diffusion equation is applied to spore dispersal within crops it is usually solved numerically and it becomes impractical to include the whole of

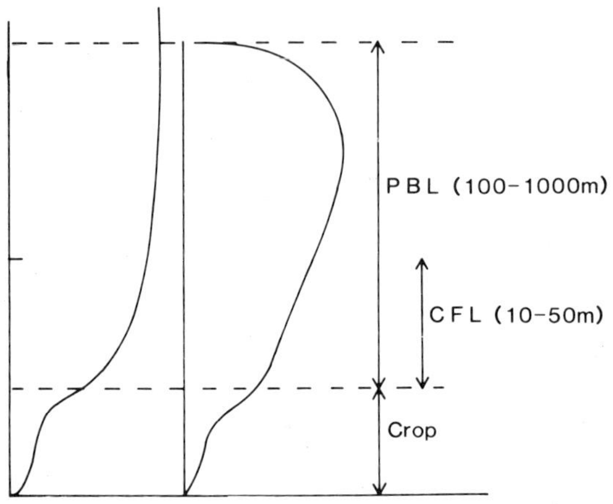

Fig. 2. Wind speed and diffusivity profiles within crops and above crops (height axis not to scale), illustrating the constant flux layer (CFL) and planetary boundary layer (PBL).

the PBL in a numerical grid. Consequently the upper boundary condition is more difficult to define. Legg and Powell (1979) and Aylor (1982) assumed that the total upward flux of spores changed linearly with height at some arbitrary height (4.5 m), whilst Aylor and Taylor (1982) assumed that the spore concentration fell to zero a few metres above the crop. However, they showed that the form of the boundary conditions had little effect on spore concentrations within the crop near a source.

For dispersal above crops, the 'ground' ($z = 0$) is taken to be the top of the crop, but for dispersal within crops $z = 0$ is the soil surface. If all particles reaching the ground were deposited, the ground would be a 'perfect sink' and the boundary condition $C = 0$ at $z = 0$ would apply. However, this is rarely the case and the flux of particles to the ground is taken to be proportional to the concentration near the surface (Chamberlain, 1975). The constant of proportionality v_d, which has units of speed, is called the deposition velocity and the appropriate boundary condition is

$$K_z \frac{\partial C}{\partial z} + v_s C = v_d C \text{ at } z = 0 \tag{10}$$

For dispersal above crops v_d describes particle deposition onto the whole crop. Legg and Powell (1979) and Aylor and Taylor (1982) assumed that deposition at ground level within a crop was predominantly by sedimentation (i.e. $v_d = v_s$) giving $K_z(\partial C/\partial z) = 0$ at $z = 0$ as the ground level boundary condition. Deposition processes are discussed in greater detail in section III,D.

3. Wind Speed (u) *and Eddy Diffusivity* (K) *Profiles*

Wind speed and eddy diffusivity vary with height and to some extent with distance. When specifying wind speed and diffusivity profiles, the atmosphere can usefully be divided into three layers; the air within the crop, the layer immediately above the crop where diffusivities change linearly with height (constant flux layer, CFL) and the PBL (Fig. 2). The CFL is generally only 10–50 m deep while the depth of the PBL may range from less than 100 m on calm clear nights to over 1000 m on hot sunny days (Monteith, 1973). For dispersal within crops both the air within the crop and the CFL must be considered. Above crops dispersal in the CFL is dominant close to the source while long distance transport is affected by the variations in K and u over the whole PBL.

Wind speed profiles within crops differ greatly from those above crops. Characteristically, wind speed decreases rapidly (large wind shear) in the top third of the canopy; in the middle of the canopy wind speed decreases slowly or may remain constant before decreasing rapidly to zero as the soil surface is approached (Monteith, 1973; Grace, 1977; Raupach and Thom, 1981) (Fig. 2). However, if the lower part of the canopy is open, as in forest, wind speed may increase slightly before falling to zero at the soil surface (Shaw, 1977; Raupach and Thom, 1981).

If wind speed measurements within a crop are not available then wind speed profiles may be estimated using an equation suggested by Cionco *et al.* (1963):

$$u(z) = u(h) \exp\{\alpha[(z/h) - 1]\} \tag{11}$$

where h is crop height and the attenuation coefficient, α, has a value between 0.3 and 3.0 depending upon crop type (Table II). Alternative profiles of similar shape have been suggested by Landsberg and James (1971) and Thom (1971). As a first approximation wind speed within a crop may be taken as a fixed fraction of the wind speed at the top of the crop [i.e. $u(z) = \beta u(h)$] and values of β for different crops are suggested in Table II.

Diffusivity, like wind speed, is smaller within crops than above crops but there is some uncertainty about the shape of the profile (Grace, 1977). It has been suggested that $K(z)$ within crops can be described by an equation similar to Eq. (11) (Lemon, 1965; Denmead, 1975; Uchijima, 1975):

$$K_z(z) = K_z(h) \exp\{\gamma[(z/h) - 1]\}$$

Table II. Within-crop wind profile parameters for different crops.[a]

Crop	α	β	Reference
Maize (immature)	2.82	0.24	Cionco (1972)
Maize	1.97	0.38	Cionco (1972)
Wheat	2.45	0.3	Cionco (1972)
Rice	1.62	0.45	Cionco (1972)
Citrus orchard	0.44	0.8	Cionco (1972)
Soybean	1.3–1.6	0.5–0.4	Baldocchi *et al.* (1983)
Beans	4.0[b]	0.14	Landsberg and James (1971)

[a] α is the wind profile attenuation coefficient, Eq. (11); β is the estimated ratio of mean within-crop wind speed to wind speed at crop height.
[b] Value estimated from published results.

with γ between 2 and 5. Thom (1975) has suggested that K_z is more or less constant in the upper two-thirds of the canopy. Legg and Price (1980) use the relationships

$$K_z(z) = K_z(h) \qquad h > z > h/3$$

$$K_z(z) = K_z(h)(3z/h) \qquad h/3 > z \geq 0$$

which may be simpler to incorporate into a dispersal model.

Wind speed and K_z profiles depend on the nature of the surface and the stability of the atmosphere. In a stable atmosphere there is little vertical mixing and turbulence and K_z are small, whilst in an unstable atmosphere convective mixing enhances turbulence and K_z can be large. The wind speed profile in the constant flux layer under neutral conditions (i.e. when the temperature profile follows the adiabatic lapse rate) is described by the logarithmic wind profile (Monteith, 1973):

$$u(z) = (u^*/k)\ln[(z - d)/z_0] \qquad (12)$$

where z is measured from the soil surface. The constant, u^* (friction velocity), is closely related to the flux of momentum and k, von Karman's constant, has an accepted value of about 0.4. The parameters, z_0, the roughness length, and d, the zero plane displacement, depend on the crop structure. For many crops the value of d is approximately $0.7h$, where h is the crop height (Monteith, 1973; Brutsaert, 1975), while z_0 usually has a value of about $0.1h$. Table III shows typical values of z_0 for different surfaces. Above crops if height, z, is measured from the top of the crop there is little error in setting $d = 0$ in Eq. (12) except close to the top of the crop.

In stable or unstable conditions the wind profile departs from Eq. (11) (Panofsky, 1963):

$$u(z) = (u^*/k)\{\ln[(z - d)/z_0] - \Psi_m[(z - d)/L]\} \qquad (13)$$

Table III. Aerodynamic characteristics of different surfaces.

Surface	Crop Height h (m)	Roughness Length z_0 (m)	Zero Plane Displacement d (m)	Wind Speed[a] (m sec^{-1})	Reference
Short-mown grass	—	0.0054	—	—	Counihan (1975)
Rough pasture	—	0.01	—	—	McIntosh and Thom (1972)
Most field crops	—	0.05–0.1	—	—	McIntosh and Thom (1972)
Grass field	0.1	0.039	—	0.42 (u^*)	Slinn (1982)
Wheat	1.3	0.045	—	0.45 (u^*)	Slinn (1982)
Wheat	1.25	0.175	0.7	—	Legg and Long (1975)
Maize	2.3	0.3	1.7	—	Hosker and Lindberg (1982)
Beans	1.11	0.063	0.9	0.084 (u^*)	Thom (1971)
		0.068	0.95	0.34 (u^*)	
Soybean	1.10	0.13	0.47	2.0	Perrier *et al.* (1972)
		0.04	0.56	4.0	
Soybean	0.8	0.064	0.47	—	Baldocchi *et al.* (1983)
	1.1	0.083	0.65	—	
Sugar beet	—	0.1	—	0.4 (u^*)	Chamberlain (1967)
Cotton	1.23	0.03	1.1	0.2 (u^*)	Bache and Unsworth (1977)
		0.12	0.92	0.5 (u^*)	
Forest	10	0.5	7.5	—	Hosker and Lindberg (1982)
	20	1.0	15.0		
Deciduous forest	—	0.25[b]	0.45[c]	6.0	Rauner (1976)
		0.15[b]	0.55[c]	3.0	
Pinus sylvestris	15.5	0.93	11.8	0.3–1.8	Jarvis *et al.* (1976)
Picea abies	27.2	7–3	19.6	0.5–2.5	Jarvis *et al.* (1976)

[a] Measured above the surface; (u^*) values refer to friction velocity.
[b] z/h.
[c] d/h.

where Ψ_m (the integrated form of the Monin–Obukov generalised momentum flux, Φ_m) is an empirical function of the dimensionless parameter $(z - d)/L$, a measure of stability (Panofsky, 1963; Yaglom, 1977). The Monin–Obukov length, L, is defined as (Monteith, 1973; Wyngaard, 1973; Thom, 1975)

$$L = -\{[\rho c_p T(u^*)^3]/kgH\}$$

where T is the mean temperature of the surface layer (°K), c_p is the specific heat of air at constant pressure, g is the gravitational constant and H is the convective heat flux. $(z - d)/L$ is large and negative in unstable layers when sensible heat flux is large, and positive in stable layers when H is negative, i.e. there is a flux of sensible heat towards the crop.

In the CFL of a neutral atmosphere eddy diffusivity values for heat (K_H), momentum (K_m) and small particles (K_z) are considered equal and proportional to height (Monteith, 1973):

$$K_z = K_m = K_H = ku^*(z - d) \tag{14}$$

However, in unstable layers eddy diffusivities are (Pasquill, 1974)

$$K_m = ku^*(z - d)/\Phi_m[(z - d)/L] \tag{15}$$

and

$$K_H = ku^*(z - d)/\Phi_H[(z - d)/L]$$

The generalised momentum and heat fluxes, Φ_m and Φ_H are also empirical functions of $(z - d)/L$ and many interpolation formulae have been proposed (Yaglom, 1977).

In highly unstable layers free convection is dominant and $\Phi_H = \Phi_m^2$ (Panofsky, 1973; Thom, 1975) and $K_H \neq K_m$. Eddy diffusivity for water vapour has been found to be close to that for heat in neutral and unstable layers (Monteith, 1973) and as a result it is often assumed that $K_z = K_H$ (Legg and Powell, 1979; van Ulden, 1978; Gryning *et al.*, 1983). However, Pasquill (1974) suggests that K_z can be taken as equal to K_m in all conditions except when the particle concentration is closely correlated with temperature.

Above the CFL wind speed increases until it reaches the geostrophic value, G, at the top of the PBL. G depends only on the horizontal pressure gradient, dP/dx (normal to the isobars), and the Coriolis parameter, f, (Tennekes and Lumley, 1972; Smith and Blackall, 1979):

$$G = (1/f\rho)(dP/dx)$$

where ρ = air density, $f = 2\Omega \sin \Phi$, Ω is the angular velocity of the earth (7.29×10^{-5} rad sec^{-1}) and Φ is the latitude. The geostrophic wind blows parallel to the isobars and in the northern hemisphere in a clockwise direction round centres of high pressure and anticlockwise round centres of low pressure. Near the ground surface, frictional drag causes a reduction in wind speed and an anticlockwise rotation in wind direction (Eckman spiral).

In many dispersal studies (Sutton, 1953; Maul, 1977; Huang, 1979; Hassid, 1983), the wind profile in the PBL is assumed to be a power law:

$$u(z) = u(z_1)(z/z_1)^n$$

where $u(z_1)$ is the wind speed at an arbitrary height z_1. The exponent n can be determined from the geostrophic wind speed and PBL depth, a:

$$n = \ln[G/u(z_1)]/\ln(a/z_1)$$

$u(z_1)$ can be estimated from Eqs. (12) or (13) if measurements are not available. Smith and Blackall (1979) suggest describing the wind profile by adding a linear term so that $u(z)$ approaches G at the top of the PBL:

$$u(z) = u'(z) + [(z - z_0)/(a - z_0)][G - u'(a)]$$

where $u'(z)$ is given by Eqs. (12) or (13).

The height of the PBL, a, is usually taken as the height of the first atmospheric temperature inversion and can be found from atmospheric temperature readings usually available from local meteorological organisations. In the absence of such measurements a can be estimated for a neutral atmosphere from u^* and the Coriolis parameter, f (Tennekes and Lumley, 1972; Schayes, 1982):

$$a = (ku^*)/f$$

Smith and Blackall (1979) suggest estimates based on sensible heat flux, H, for unstable and stable layers:

$$a = 21{,}500\, u^*(H)^{-0.3} \qquad \text{stable}$$

$$a = a\,(\text{neutral}) + a' \qquad \text{unstable}$$

where

$$a' = 2.3H + 1.25 \times 10^{-3}H^2 + 1.63 \times 10^{-5}H^3$$

Few measurements of diffusivity have been made over the full depth of the PBL; thus the form of K_z is uncertain. Yokahama *et al.* (1977) showed that in a neutral atmosphere K_z could be estimated:

$$K_z = cu^*z(1 - z/a)$$

where c is an empirical constant. This equation estimates the profile of K_z with a maximum value at $z = 0.5a$ and K_z approaching zero at $z = a$. The constant c can be found from K_z at an arbitrary height z_1 estimated from Eq. (21) or (22). Wippermann (1972), using a PBL model, estimated K_z profiles:

$$K_z = zku^* \exp[-c(fz/ku^*)^{0.764}]$$

where the constant c depends on atmospheric stability. Schayes (1982) estimated c:

$$c = \exp(0.264 + 0.0162\mu + 0.00396\mu^2)$$

where $\mu = a/L + 50$ and found reasonable agreement with experimental measurement.

Measurement of wind speed and eddy diffusivity profiles for dispersal models is not easy. K_z cannot yet be measured directly for particles and must be estimated from K_m, K_H or K for some nonreacting gas such as nitrous oxide (Legg, 1975; Legg and Long, 1975) or CO_2 (Johnson *et al.*, 1976).

Above crops K_m and K_H can be calculated using Eq. (14) or (15) if u^*, d and L are known. These may be calculated, using iterative procedures, from measured wind profiles and estimates of the convective heat flux H (Legg and Long, 1975; Berkowicz and Prahm, 1982). Schayes (1982) and Holtslag and Van Ulden (1983) have suggested methods for estimating u^*, and H (hence L) from standard meteorological measurements. Values of L corresponding to the widely used Pasquill atmospheric stability classes (Pasquill, 1962), which are readily estimated from routine meteorological observations, are given in Table IV.

K_m and K_H can also be derived from measurements of momentum or heat flux:

$$F_z = -K_z(\partial C/\partial z)$$

where $C = \rho u$ for momentum and $\rho c_p T$ for heat. Heat and water vapour fluxes above and within crops can be estimated from measurements of net radiation, soil heat flux and profiles of temperature and water vapour by considering the energy balance of the crop (the Bowen ratio method) (Monteith, 1973). F_z can also be determined directly by time averaging the product of instantaneous measurements of fluctuations in vertical wind speed ($w' = \bar{w} - w$) and concentration ($C' = \bar{C} - C$) (Sutton, 1953):

$$F_z = \overline{w'C'}$$

The 'eddy correlation' method, as it is often called, requires fast, sensitive instruments and has been little used until recently (Bottemanne, 1979; Champagne *et al.*, 1977; Maitani, 1977; Shaw, 1977; Motha *et al.*, 1980).

The above discussion assumes that particles behave as fluid elements with eddy diffusivities the same as those of diffusing gases. Heavy particles may not behave as fluid elements because of their inertia and because they can 'fall

Table IV. Atmospheric stability classes (Pasquill, 1962).

A.	Class	Description	Monin–Obukov Length,[a] L (m)
	A	Highly unstable	−2 to −3
	B	Unstable	−4 to −5
	C	Slightly unstable	−12 to −15
	D	Neutral	∞
	E	Slightly stable	35 to 75
	F	Stable	8 to 35

B. Determination of stability class from meteorological observation

Surface Wind Speed ($m\ sec^{-1}$)	Daytime Solar Radiation[b] Strong	Moderate	Slight	Overcase	Night ≤3/8 Cloud	Thinly Overcast or ≤4/8 Cloud	Overcast
<2	A	A–B	B	D	—	—	D
2–3	A–B	B	C	D	E	F	D
3–5	B	B–C	C	D	D	E	D
5–6	C	C–D	D	D	D	D	D
>6	C	D	D	D	D	D	D

[a] For short grass surface, Pasquill and Smith (1971).
[b] Strong—sunny midday midsummer in England (> 600 $W\ m^{-2}$); moderate—average summer in England (300–600 $W\ m^{-2}$); slight—sunny midday winter in England (< 300 $W\ m^{-2}$).

out' of transporting eddies. However for particles with fall speeds of less than about 2 m sec^{-1} inertial effects can be neglected (Smith, 1959; Csanady, 1963) and that if v_s is less than the fluctuations in vertical wind speed K can be taken equal to for gases (Csanady, 1963). Within crops this condition is likely to be satisfied if the effective particle diameter is less than 40 μm (Legg, 1983a). Above crops where eddies are larger the theory may be applied to larger particles. Thus K theory is applicable to the dispersal of many airborne pathogen propagules but may be inappropriate for larger splash droplets which have an appreciable fall speed. However, Csanady (1963) and Smith and Blackall (1979) suggest methods of correcting K for such 'heavy' particles.

4. *The Source and Sink Term* (S)

For dispersal above crops deposition of spores is accounted for by the boundary conditions at $z = 0$ [Eq. (14)] and S is zero at all points except at spore sources where it represents the rate of change in concentration due to spore release. Within crops, S represents the net change in concentration which depends on the rate of spore release and the loss of spores by deposition onto plant surfaces. Spore deposition rate is proportional to the local concentration C. Thus, in the absence of a local source:

$$S = -rC \tag{16}$$

where r, the deposition rate constant (similar to deposition velocity but with units $\sec^{-1}$), is a function of wind speed, spore size and canopy structure; r, which includes deposition by sedimentation and by impaction, is discussed in section III,D.

5. Analytical Solutions to the Diffusion Equation

Because the forms of eddy diffusivity and wind speed profiles in the atmosphere are complex, it is frequently impossible to find analytical solutions to the diffusion equation and numerical methods must be used. However, analytical solutions offer several advantages: they can be incorporated into disease development models more easily; they often illustrate the dependence of diffusion on particular parameters and the effects of multiple sources can be assessed more easily.

Analytical solutions to simplified atmospheric diffusion equations have been reviewed by Pasquill (1974), Fisher (1979) and Hanna (1982). Additionally, Carslaw and Jaeger (1959) give solutions to the diffusion equation for heat conduction in solids which can be applied to atmospheric diffusion. Even for simple forms of K and u, solutions are generally complex and require iterative procedures to evaluate them. Analytical solutions may be particularly useful for describing long-distance transport of spores. For example, Tang (1969) has investigated the simple case of particles dispersing from an elevated line source in an unbounded atmosphere where K_z and u are independent of height.

Spore dispersal within a homogeneous canopy can be described by an analytical solution developed by Llewellyn (1983) for diffusion of particles undergoing a first-order chemical reaction from a point source. He assumed that particles were removed from the diffusing cloud at a rate proportional to their concentration [i.e. equivalent to S defined by Eq. (16)], that K_x, K_y, K_z, u and v were constant within a mixing layer of depth a and that $w = 0$. These conditions are equivalent to those within plant canopies if a is taken as the crop height and u as the wind speed at mid-canopy height. Loss of spores from the crop is represented by the leakage boundary condition [Eq. (9)] which implies that the flux of spores out of the crop is proportional to the concentration near the top. This assumption is unrealistic at large distances from the source where there may be a net downward flux of spores into the crop due to spores re-entering the canopy. However, Llewellyn's solution may be used to estimate deposition gradients close to a source and to investigate the relative importance of diffusivity, wind speed, deposition to vegetation and deposition to the ground on pathogen dispersal.

6. Numerical Solutions to the Diffusion Equation

With the advent of fast computers it has become possible to use numerical techniques to solve the diffusion equation with realistic boundary conditions, eddy diffusivity and wind speed profiles. Although such techniques offer great flexibility, care must be taken to avoid numerical difficulties such as instabilities. There has been much work on this problem and several techniques have been reviewed by Long and Pepper (1976) and Chock and Dunker (1983). It is not our intention to describe the various computational schemes available, and we refer the reader to the original papers. Legg and Powell (1979), Aylor (1982) and Aylor and Taylor (1982) have used a specific technique, the finite difference method, to solve the two-dimensional form of the diffusion equation applied to spore dispersal within crop canopies.

B. Statistical (Gaussian Plume) Models

The statistical or Gaussian plume model of turbulent dispersal, perhaps the most widely used method for predicting concentrations of atmospheric pollutants, can readily be applied to spore dispersal in some circumstances. The model assumes that the average distributions of particles across the height and width of a plume downwind from a continuous point source can be described by Gaussian curves (Pasquill, 1962; Turner, 1970; Hanna, 1982) (Fig. 3). The standard deviations σ_z and σ_y of the concentration distribution in the z (vertical) and y (cross-wind) directions vary with downwind distance, x, and the total flux of particles across a vertical plane at right angles to the mean wind (x) direction is constant, i.e. there is no loss of particles from the plume. The mean particle concentration downwind of point source at height H is given by

$$C(x,y,z) = \frac{Q}{u}\frac{\exp(-y^2/2\sigma_y^2)}{2\pi\sigma_z\sigma_y}\cdot\left\{\exp\left[-\frac{(H-z)^2}{2\sigma_z^2}\right]+\exp\left[-\frac{(H+z)^2}{2\sigma_z^2}\right]\right\} \quad (17)$$

where Q is the source strength and u the mean wind speed at release height. The calculated concentrations are equivalent to measured concentrations averaged over a time similar to the time of travel of the particles from the source to the point of measurement (Pasquill, 1962).

Concentration downwind of a continuous line source emitting Q particles per unit time per unit length is

$$C(x,z) = \frac{Q}{u}\frac{1}{\sqrt{2\pi}\sigma_z}\cdot\left\{\exp\left[-\frac{(H-z)^2}{2\sigma_z^2}\right]+\exp\left[-\frac{(H+z)^2}{2\sigma_z^2}\right]\right\} \quad (18)$$

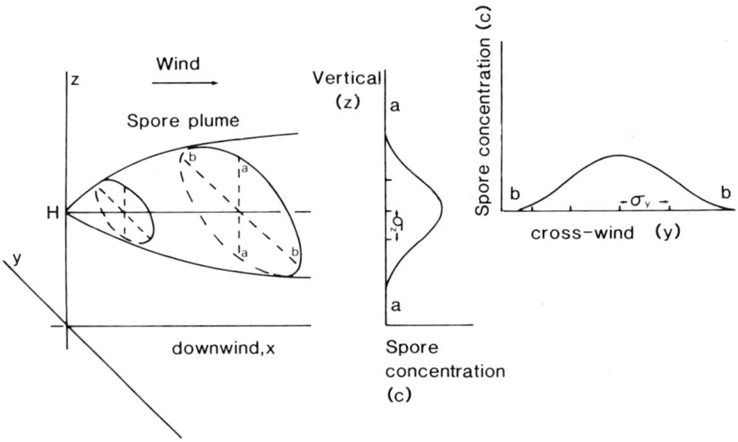

Fig. 3. Spore dispersal downwind of a point source at height *H*. The *x* axis represents the mean wind direction, the *y* axis the cross-wind direction and the *z* axis the vertical direction. Gaussian distributions of spore concentration *C* in the vertical (*a–a*) and cross-wind (*b–b*) directions, with standard deviations σ_z and σ_y, are illustrated. (See text, section III,B.)

For dispersal of plant pathogens above crops, *z* is measured with respect to crop height and $H = 0$. For ground level concentration, $C(x,0)$, Eqs. (17) and (18) simplify to

$$C(x,0) = \frac{Q}{u} \cdot \frac{\exp(-y^2/\sigma_y^2)}{\pi \sigma_z \sigma_y} \tag{19}$$

$$C(x,0) = Q(2/\pi)^{0.5}/u\sigma_z \tag{20}$$

There have been many experimental determinations of the variation in σ_z and σ_y with distance *x* and with atmospheric stability. Generally, the less turbulent the atmosphere the more slowly a plume expands with distance downwind. There have been several recent reviews of the formulation of σ_z and σ_y in the atmosphere (Pasquill, 1976; Gifford, 1976; Hanna *et al.*, 1977; Sedefian and Bennett, 1980; Mitchell, 1982). However, much use is made of the Pasquill atmospheric stability classes (Table IV) and analytical approximations of σ_z and σ_y for each class are given in Table V.

There is little information available on the variation in σ_z and σ_y with downwind distance within crops. Aylor (1976) discussed the determination of σ values for an expanding plume moving downwind within a forest canopy using measured horizontal wind speed power spectra. Direct measurements of turbulence can be used to estimate σ_y and σ_z (Hanna, 1982) and turbulence observations within crops may be used to derive appropriate σ_z and σ_y values.

Table V. Analytical representations of σ_z and σ_y in the planetary boundary layer (Briggs, 1974).[a]

Stability Class	σ_y (m)	σ_z (m)
A	$0.22x\ (1 + 0.0001x)^{-0.5}$	$0.20x$
B	$0.16x\ (1 + 0.0001x)^{-0.5}$	$0.12x$
C	$0.11x\ (1 + 0.0001x)^{-0.5}$	$0.08x\ (1 + 0.002x)^{-0.5}$
D	$0.08x\ (1 + 0.0001x)^{-0.5}$	$0.06x\ (1 + 0.0015x)^{-0.5}$
E	$0.06x\ (1 + 0.0001x)^{-0.5}$	$0.03x\ (1 + 0.0003x)^{-1}$
F	$0.04x\ (1 + 0.0001x)^{-0.5}$	$0.016x\ (1 + 0.0003x)^{-1}$

[a] Valid for $100 < x < 10{,}000$ m.

Sutton (1953) suggested that the σ values of a diffusing plume may be expressed as

$$\sigma = Ax^m \tag{21}$$

where A and m are constants. Gregory *et al.* (1961) showed that Eq. (21) fitted data on the dispersal of *Lycopodium* spores released over grass, with m between 1.2 and 1.75. Reddi (1975) found similar values of m for vertical and horizontal spread of *Lycopodium* and *Podaxis* spores released 1 m above open ground.

Equations (17) to (20) assume that deposition of particles is negligible and thus overestimate the concentration of spores downwind of a source. Deposition can be partially accounted for by reducing the source term, Q, downwind of the source (Gregory, 1945; Smith and Singer, 1966; Van der Hoven, 1968; Pasquill, 1974). In this 'source depletion' model the shape of the concentration profile remains Gaussian and unchanged by the deposition process. If the deposition flux to the ground, F, is described by a deposition velocity v_d [Eq. (10)] then deposition rate, R, is

$$R(x,y) = C(x,y,z') \cdot v_d$$

where z' is a reference height and C the concentration. Then, according to Horst (1977), Q in Eq. (17) can be replaced by

$$Q(x) = Q \cdot \exp\left[-\int_0^x \frac{v_d}{u} D(x')\,dx'\right] \tag{22}$$

where

$$D(x) = \frac{1}{\sqrt{2\pi}\sigma_z} \cdot \left\{\exp\left[-\frac{(H-z')^2}{2\sigma_z^2}\right] + \exp\left[-\frac{(H+z')^2}{2\sigma_z^2}\right]\right\}$$

The value of the integral in Eq. (22) depends on the form of σ_z and may be determined numerically. The source depletion approximation implies that the particles are well mixed within the plume and ignores the effects of deposition on the concentration profile. As a result deposition and surface concentration are overestimated at distances greater than 100 m downwind especially if v_d is large (Horst, 1977).

Within a crop, particles are removed from all parts of the plume and the assumption that the plume shape remains constant may be valid, especially close to a source when most of the plume remains within the canopy. Deposition per unit volume is described by Eq. (16) where r, the deposition rate constant, can be taken as a function of height only. By considering the conservation of particle numbers (Pasquill, 1974; Horst, 1977), concentration can be calculated replacing Q in Eq. (17) by

$$Q(x) = Q \exp\left[-\int_0^x \int_0^\infty \frac{r}{u} D(x',z)\mathrm{d}z\, \mathrm{d}x'\right]$$

The value of the integrals in this equation depend on the forms of r, u and σ_z. If r and u are independent of z (a homogeneous crop with uniform wind profile) $Q(x)$ simplifies to

$$Q(x) = Q \exp(-rx/u)$$

Equations (17) to (20), incorporating source strength modifications to account for deposition, allow prediction of particle concentration downwind from point and line sources. Spore deposition rates and hence deposition gradients can be estimated from concentrations using appropriate deposition velocities.

Gaussian plume models cannot accurately predict spore concentrations and deposition rates in complex situations such as nonhomogeneous crops or boundary layers with large wind shear. However, they are relatively simple to use with easily acquired meteorological information (Tables IV and V). Thus when ease of use is more important than accuracy, such models may be useful. Indeed they have been used for many years in air pollution control studies and in practical air quality control projects (Turner, 1970).

C. Random Walk Models

The gradient transfer theory and statistical dispersal models can predict dispersal of plant pathogens above crops downwind of sources with acceptable accuracy under most atmospheric conditions. However, although they have been used to predict dispersal within crops, they have several limitations (Raupach and Thom, 1981). Additionally, gradient transfer and Gaussian plume theory implicitly assume that the particle release rate is independent of wind speed. This may be true for actively released spores such as those of *Sphaerotheca mors-uvae* (Ingold, 1971) or for spores that are only loosely held on the host leaf surface. However, spores of many plant pathogens require considerable force to remove them from leaves (Aylor and Parlange, 1975; Aylor, 1975) and may be liberated only when the wind speed is above a given minimum value, i.e. when wind gusts penetrate the canopy (Aylor, 1978). Such intermittent release may affect spore deposition patterns by increasing the efficiency of deposition close to the source (Aylor, 1978; Aylor *et al.*, 1981).

Random walk models simulate the trajectories of individual spores or air parcels and concentration and deposition patterns are built up from many individually calculated particle trajectories. The trajectories of individual fluid elements are simulated as a pseudo-random walk from a knowledge of the turbulence statistics of the fluid flow (Hall, 1975; Hanna, 1982; Ley, 1982). Spores or particles are assumed to follow the same path as the air parcels but with an additional downward velocity equal to their settling velocity (Legg, 1983a). Such models have been applied to agricultural spray drift above crops (Thompson and Ley, 1982; Thompson, 1983) and to spore dispersal within crop (Legg, 1983a). Air parcel trajectories are represented as a connected series of discrete displacements determined partly by a correlation between successive velocities and partly by a random component to account for turbulent fluctuations. Such calculations can be treated as a Markov chain process. For example, for homogeneous turbulence the sequence for vertical velocity fluctuations, w, is (Legg, 1983a)

$$w_{n+1} = \alpha_w w_n + \beta_w \sigma_w \xi \tag{23}$$

where w_n and w_{n+1} are the vertical velocities during steps n and $n + 1$, σ_w is the standard deviation of w, ξ is a random number taken from a distribution with mean zero and unit variance, and α_w and β_w are constants. For the velocity autocorrelation time scale (i.e. the velocity 'memory' of the element) to be correct and equal to the Lagrangian time scale τ_w, (a measure of the life time of an average eddy), Hall (1975) showed that α_w and β_w take the following form:

$$\alpha_w = \exp(-\Delta t/\tau_w)$$

$$\beta_w = (1 - \alpha_w^2)^{0.5}$$

where Δt is the time step chosen. Similar expressions can be written for the along-wind, u', and cross-wind, v', components of particle velocity fluctuation (Legg, 1983b). The along-wind velocity, u, is found by adding the mean value to u', i.e. $u = \bar{u} + u'$.

Turbulence within crops is not homogeneous; e.g. standard deviations in horizontal and vertical wind speed, σ_u and σ_w, and shear stress, $\overline{u'w'}$, are functions of height. Wilson *et al.* (1981) showed that Eq. (23) cannot be used in these circumstances and Legg and Raupach (1982) accounted for this by expressing the sequence for vertical fluctuations as

$$w_{n+1} = \alpha_w w_n + \beta_w \sigma_w \xi + (1 - \alpha_w)\tau_w \partial\sigma_w^2/\partial z$$

and fluctuations along the mean wind direction as

$$u'_{n+1} = \alpha_u u'_n + \beta_u \sigma_u \zeta + (1 - \alpha_u)\tau_u \partial(\overline{u'w'})/\partial z$$

where ζ is a pseudo-random number chosen from a Gaussian distribution so that (Legg, 1983b)

$$\overline{\xi\zeta} = \overline{u'w'}(1 - \alpha_u\alpha_w)/(\sigma_u\sigma_w\beta_u\beta_w)$$

to maintain correlation between u' and w' (the overbars indicate mean values).

To apply the above equations to spore trajectories the spore settling velocity must be added to the expression for w (Thompson and Ley, 1982; Thompson, 1983). As the theory assumes that the spores are small enough to follow the fluctuations in wind speed exactly and that they do not 'fall out' of the eddies in which they are travelling before the eddies decay, the same restrictions on particle size apply as for K theory (section III,A).

In order to use random walk models, detailed information on turbulence statistics and Lagrangian time scales is required. Above crops such information may be obtained by measurement or from theories of atmospheric turbulence (Hanna, 1982; Ley, 1982). Unfortunately, Lagrangian time scales are difficult to measure and little information on the required turbulence statistics and time scales within crops is available. Thus there remains much theoretical and experimental work to be done before random walk models can reach their full potential.

Deposition processes can be incorporated into random walk models by determining the probability that spores will impact or sediment onto foliage elements during each time step (Legg, 1983a). Sedimentation depends only on the spore fall speed and the spatial distribution of foliage while impaction is also a function of wind speed $(u^2 + v^2 + w^2)^{0.5}$ and particle size (section III,D).

Although random walk models require further development they offer several advantages over conventional dispersal theories. Random walk theory can be used to describe dispersal in highly turbulent, heterogeneous atmospheric conditions, is valid at any distance from the source and can allow for different mechanisms of spore release. Random walk simulation is, therefore, a most useful tool for studying dispersal of a wide range of particles under different meteorological conditions to gain a better understanding of the interaction between wind, crop and disease.

D. Deposition Processes

Deposition of particles from the atmosphere onto a crop can be calculated using the deposition velocity, v_d [Chamberlain, 1975; Eq. (10) section III,A]; deposition rate (particles per unit surface area per unit time) D is

$$D = v_d C(z')$$

where $C(z')$ is the concentration at some reference height, z', close to the crop surface. There have been many investigations of particle deposition, both onto plants in wind tunnels and onto crops in the field, and Table VI gives some measured values of v_d. Deposition velocity depends on several factors: particle size and structure, which determine the settling velocity of the particle and whether it is retained on the foliage after deposition; crop structure and leaf density, which affect deposition processes; wind speed and turbulence, which influence transfer into crops and the impaction and sedimentation of particles onto plants. Generally v_d is greater than the particle settling velocity, v_s, except at very low wind speeds or when spores 'bounce off' plant surfaces (Chamberlain, 1975). The ratio v_d/v_s is usually between 2 and 5 (Table VI) and increases with friction velocity, u^* (Chamberlain and Chadwick, 1972; Chamberlain, 1975; Callander and Unsworth, 1983).

The difference between v_d and v_s is due to deposition of particles onto the crop by turbulent diffusion and can be written

$$v_d = v_s + v_t$$

where v_t represents the component due to turbulent deposition $\{v_t \equiv [(K/C)(\partial C/\partial z)_{z=z'}]\}$. Making the assumption that the eddy diffusivity for particles is related to that for momentum then the upper limit for v_t is v_m, the deposition velocity for momentum:

$$v_m = u^{*2}/u(z')$$

Dollard and Unsworth (1983) found that the mean value of v_t/v_m for fog droplets ($\sim 20\ \mu m$) depositing onto grass was 0.97 ± 0.36 when $u(z')$ was greater than 2 m sec^{-1}. Callender and Unsworth (1983) found that v_d for 30- to 60-μm droplets depositing onto wheat could be estimated as

$$v_d = v_s + 0.1u^*$$

which agreed with measurements of *Lycopodium* spore deposition onto a 'wet crop' (little bounce off) by Chamberlain and Chadwick (1972). Legg and price (1980) showed theoretically that v_t could make a significant contribution to v_d even when deposition onto the crop was by sedimentation only.

Table VI. Published values of deposition velocity, v_d, and settling velocity, v_s, for different particles.

Particle	Size (μm)	Surface	v_d (cm sec^{-1})	v_s (cm sec^{-1})	Wind Speed (u) or Friction Velocity (u^*) (m sec^{-1})	Reference Height z' (m)	Remarks	Reference
Lycopodium spores	32	Short rough grass	1.0–10.0 (mean 4.0)	—	0.4–2.64 (u)	0.25–1.0	10 m from source	Gregory *et al.* (1961)
Lycopodium spores	32	Short grass	0.7	1.9	0.12 (u^*) 0.53 (u^*)	—	—	Chamberlain (1967)
Lycopodium spores	32	Short grass	3.5	1.75	1.1–3.6 (u)	0.25	10 m from source	Reddi (1976)
Lycopodium spores	32	Barley	1.4–2.1	1.8	—	—	—	Legg and Bainbridge (1978)
Ganoderma applanatum spores	10.5 × 6.5	Rough grass	0.08–0.34	0.22	—	—	5 m from source	Gregory *et al.* (1961)
Fomes annosus spores	5	Forest floor	0.016–1.6	—	0.05–2.2 (u)	—	—	Edmonds and Driver (1974)
Erysiphe graminis spores	22 × 15	Barley	0.7–4.3	1.2	—	—	Naturally released spores	Legg and Bainbridge (1978)

Podaxis spores	14 × 11	Short grass	0.68	—	1.1–3.6 (u)	0.25	Mean 10 m from source	Reddi (1976)
Phleum pratense (timothy) pollen	25–35	Rough ground	11.5 ± 2.2	3.3	—	1.5	36 m from area source	Raynor *et al.* (1970)
Zea mays (maize) pollen	90–100	Rough ground	63 ± 25	20	—	1.5	32 m from source	Raynor *et al.* (1970)
Ricinus communis pollen	24 × 38	Short grass	4.0	—	—	0.5	30 m from line source	Raynor *et al.* (1973)
Ambrosia trifida pollen	20		3.0	1.54	—	0.5	30 m from line source	Raynor *et al.* (1973)
Kachia scoparia pollen	30		10.0	—	—	0.5	30 m from line source	Raynor *et al.* (1973)
Droplets	30–60	Bare soil	20 ± 10	7.3	0.5 (u^*)	0.25	Down-wind from line source	Callander and Unsworth (1983)
		Young wheat (0.08 m)	17 ± 4	8.0	0.56 (u^*)	0.25		
		Wheat (0.8 m)	6.8 ± 0.9	4.9	0.37 (u^*)	0.8		
		Wheat (1.0 m)	21 ± 5	10.5	0.39 (u^*)	1.0		
Fog droplets	20	Grass	24–70	17–23	0.24–0.40 (u^*)	1.0	—	Dollard and Unsworth (1983)

Bache and Uk (1975), Bache (1979a,b), Legg and Price (1980) and Slinn (1982) have presented theories to relate particle deposition velocities to crop structure and turbulence. Slinn (1982) considered the case where deposition was predominantly by inertial impaction, i.e. for crops with largely erect foliage. Bache (1979a,b) allowed for both impaction and sedimentation in his model and an example of its use is given in Bache (1981). All three theoretical investigations are complex and are not readily summarised and we leave the reader to consult the original papers for details.

Within crops, deposition by gravitational settling and by inertial impaction are often treated as separate processes (Legg and Powell, 1979; Bache, 1979a,b; Aylor, 1982). When deposition rate per unit volume, S, is expressed as $-rC$ the coefficient r contains effects of both. Partition can be achieved by assuming that sedimentation predominates on horizontal leaf elements and impaction predominates on vertical elements (Bache, 1979a,b; Legg and Powell, 1979). This assumption is not strictly valid because the variable boundary layer, which has a wind speed component perpendicular to the surface, above a leaf may affect sedimentation. Non-vertical or non-horizontal leaves are resolved into areas projected onto the x (horizontal) and z (vertical) directions to determine the proportions of total deposition which are by sedimentation and by impaction. Measurements of the influence of orientation on deposition suggest that when sedimentation is dominant (i.e. low wind speeds) this treatment is acceptable (Bache, 1979a). When impaction is dominant (i.e. at high wind speeds) measurement suggests that the partitioning of deposition into sedimentation and impaction in this simple manner may not be adequate (Gregory and Stedman, 1953).

If f_z and f_x are the fractions of the plant area density (area/unit volume), A, projected vertically (z) and horizontally (x) then deposition by sedimentation is

$$S_s = C \cdot f_z A v_s \tag{24}$$

where C is the concentration and v_s is the settling velocity. A and f_z are functions of height and depend on crop density and structure. Monsi *et al.* (1973) have examined the distributions of leaf angle for several crops and from those distributions f_z and f_x can be estimated. This treatment ignores 'enhanced sedimentation' onto horizontal surfaces due to impaction of spores traveling in eddies with large vertical velocity components.

Deposition by impaction can be regarded as the sum of the deposition onto various plant elements:

$$S_i = S_i \text{ (leaves)} + S_i \text{ (stems)} + S_i \text{ (heads)} \ldots \text{etc.}$$

Each S_i is found from

$$S_i = C \cdot u \cdot f_x A \cdot E \tag{25}$$

where A and f_x are the area densities and fraction of the vertical component for each element. E, the efficiency of impaction, is defined as

$$E = \frac{\text{rate at which particles impact upon a surface}}{\text{rate at which particles pass through the same area as the surface projected horizontally}}$$

and is a function of wind speed, particle size and shape, and plant element size. In laminar flow the impaction efficiency of a particle of diameter d, is a function of its Stokes' number, St:

$$\mathrm{St} = v_s u/(g \cdot l)$$

where l is a characteristic length of the plant element (e.g. diameter of stems or width of leaves) and g is acceleration due to gravity. For solid spores which obey Stokes' Law (diameter $\leq 75\ \mu$m) v_s can be calculated (Chamberlain, 1975):

$$v_s = (d^2 g \rho_0)/(18 v \rho) \qquad (26)$$

where ρ_0 is the density of the sphere, ρ is the density of air (1.22 kg m^{-3} at 15°C) and v is the kinematic viscosity of air (1.47×10^{-5} m^2 sec^{-1} at 15°C). For non-spherical particles a correction factor must be applied to Eq. (26) (Fuchs, 1964; Chamberlain, 1975) and for larger particles v_s can be found from experimentally derived drag coefficients (Fuchs, 1964). Gregory (1973) presents measured values for v_s for many fungal spores and pollens. There have been several theoretical and experimental determinations of E for different values of St (Chamberlain, 1975); however, the experimental measurements of deposition onto cylinders and ribbons by May and Clifford (1967) are widely used. Aylor (1982) gives the following functional representation of their results:

$$E = 0.86/(1 + 0.442\ \mathrm{St}^{-1.967}) \qquad (27)$$

This equation assumes that all particles impacting onto a surface are retained. For many spores, e.g. those of the powdery mildews, this is probably the case. However, some spores, e.g. those of *Lycopodium*, may bounce off dry surfaces (Chamberlain and Chadwick, 1972). Bounce off can be allowed for by modifying the impaction efficiency; unfortunately there is little information on values of corrections required.

Equation (25) assumes that the average impaction efficiency can be calculated from the mean wind speed, u. This is not necessarily the case as impaction efficiency is a non-linear function of wind speed and wind speed frequency distributions within crops are highly skewed (Shaw *et al.*, 1979; McCartney *et al.*, 1983). Aylor *et al.* (1981) showed that the difference between impaction efficiency calculated from the mean wind speed and impaction efficiency estimated from the wind speed frequency distribution was small for spores that are released at all wind speeds. However, for spores released only in gusts impaction efficiencies may be enhanced, particularly close to the source where the mean spore velocity will be greater than the mean wind speed (Aylor *et al.*, 1981; McCartney *et al.*, 1983). The appropriate impaction efficiency for spores released in gusts is defined by Aylor *et al.* (1981):

$$E = \int_{u_t}^{\infty} E(u)\, f(u)\, \mathrm{d}u$$

where u_t is the threshold wind speed for spore release, $E(u)$ is the impaction efficiency at u and $f(u)$ is the wind speed frequency distribution. Efficiencies calculated in this way can be several times larger than those estimated from the mean wind speed, especially for large u_t and small Stokes' number (Aylor *et al.* (1981). Aylor (1982) calculated E in this way in a model describing the dispersal of spores released in gusts.

The relationship between impaction efficiencies and Stokes' number in Eq. (27) was derived from measurements made in low turbulence air flow. Within crops mean wind speeds are usually low but turbulence (wind speed fluctuations) can be high, even leading to flow reversal, and deposition processes may not be adequately described by relationships derived from low turbulence measurements. Recent wind tunnel measurements in turbulent flow by Jones (1983) have shown that deposition is enhanced by turbulence at low Stokes' numbers. This may explain why the deposition of *E. graminis* spores onto barley at low Stokes' number was greater than deposition predicted by Eq. (27) (Bainbridge and Stedman, 1979). The deposition rate coefficient, r, is then found by combining Eqs. (24) and (25):

$$r = f_z A v_s + u[(f_x AE)_{\text{leaves}} + \cdots]$$

IV. CONCLUSIONS

Empirical models of spore dispersal have proved extremely useful to epidemiologists for identifying sources (Gregory and Lacey, 1964), for determining the relative importance of primary and secondary inoculum (Rowe and Powelson, 1973; Johnson and Powelson, 1983) and for choosing the minimum isolation distance from a potential source at which to plant a susceptible crop. They have helped in the assessment of cultivar mixtures (Jeger *et al.*, 1983) and multilines (Fried *et al.*, 1979) for reducing epidemic development and in the interpretation of small plot experiments with air-borne pathogens where interplot interference occurs (Jenkyn *et al.*, 1979; Paysour and Fry, 1983). Unfortunately there has been no systematic analysis to relate empirical dispersal parameters (gradients) to spore type, crop structure and weather.

These relatively simple models can easily be incorporated into more general models of disease epidemic development in time and space (Jeger, 1983; McCartney and Fitt, 1985) for use in disease forecasting and control programmes. However, they provide little information about the mechanisms of spore dispersal: spore removal, transport and deposition.

It is possible, in principle, to construct physical models of spore dispersal which are generally applicable in many different situations. However, although such physical models are widely used in air pollution studies and have been used to predict the dispersal of spores (Gregory, 1973; Aylor, 1978; Legg and Powell, 1979; Legg, 1983a) and bacteria (Graham *et al.*, 1977), their application in plant pathology is still in its infancy. As physical models are generally complex their direct use in epidemiological studies is limited. However the results produced, suitably formulated, can be incorporated into general epidemic models.

The further development of spore dispersal models for use in plant pathogen epidemiology is restricted by a lack of accurate experimental measurements with which to test them. For example, there is little information on spore deposition in turbulent flow and on air turbulence characteristics within crops. There have been few experimental studies of spore dispersal in wind tunnels, where environmental conditions can be accurately controlled, which might provide some of the information about spore dispersal. Spore dispersal gradients measured in the field have rarely been accompanied by sufficient (if any) meteorological observations or crop structure measurements to enable reliable estimates of dispersal or deposition parameters to be made. Accurate meteorological instruments, essential for recording the short-lived weather changes important in spore dispersal, and computer-controlled data loggers are now available (Sutton *et al.*, 1984) and it should now be possible to make the necessary meteorological measurements.

Dispersal is an important stage in pathogen life cycles and it is widely recognised that lack of information about dispersal is limiting the understanding of epidemic development. Theories to estimate pathogen dispersal and deposition now exist, although there is still much to be learned before they can be applied fully. Thus we hope that in the next decade a substantial effort will be made in the field and in the laboratory to gather the information necessary for their application to plant disease epidemiology.

REFERENCES

Aylor, D. E. (1975). *Plant Physiol.* **55**, 99–101.

Aylor, D. E. (1976). *In* "Perspectives in Forest Entomology" (J. F. Anderson and H. K. Kaya, eds.), pp. 177–188. Academic Press, New York.

Aylor, D. E. (1978). *In* "Plant Disease: an Advanced Treatise" (J. G. Horsfall and E. B. Cowling, eds.) Vol. 2, pp. 159–180. Academic Press, New York.
Aylor, D. E. (1982). *Agric. Meteorol.* **26**, 215–219.
Aylor, D. E. and Parlange, J.-Y. (1975). *Plant Physiol.* **56**, 97–99.
Aylor, D. E. and Taylor, G. S. (1982). *Proc. Natl. Acad. Sci., U.S.A.* **79**, 697–700.
Aylor, D. E., McCartney, H. A. and Bainbridge, A. (1981). *J. Appl. Meteorol.* **20**, 1212–1221.
Aylor, D. E., Taylor, G. S. and Raynor, G. S. (1982). *Agric. Meteorol.* **27**, 217–232.
Bache, D. H. (1979a). *Atmos. Environ.* **13**, 1257–1262.
Bache, D. H. (1979b). *Atmos. Environ.* **13**, 1681–1689.
Bache, D. H. (1981). *Atmos. Environ.* **15**, 1759–1761.
Bache, D. H. and Uk, S. (1975). *Agric. Meteorol.* **15**, 371–377.
Bache, D. H. and Unsworth, M. H. (1977). *Q. J. R. Meteorol. Soc.* **103**, 121–134.
Bainbridge, A. and Stedman, O. J. (1979). *Ann. Appl. Biol.* **91**, 187–198.
Baldocchi, D. D., Verma, S. B. and Rosenberg, N. J. (1983). *Boundary Layer Meteorol.* **25**, 43–54.
Berkowicz, R. and Prahm, L. P. (1982). *Atmos. Environ.* **16**, 2809–2819.
Bottemanne, F. A. (1979). *Agric. Meteorol.* **20**, 397–410.
Briggs, G. A. (1974). *In* "Environmental Research Laboratories, Air Resources Atmospheric Turbulence and Diffusion Laboratory 1973 Annual Report." U.S. Atomic Energy Comm. Rep. ATDL-106. Natl. Oceanic and Atmos. Admin. Washington, D.C.
Brutsaert, W. (1975). *J. Meteorol. Soc. Jpn.* **53**, 96–97.
Buchanan, T. S. and Kimmey, J. W. (1938). *J. Agric. Res.* **56**, 9–30.
Callander, B. A. and Unsworth, M. H. (1983). *Agric. Meteorol.* **28**, 109–127.
Cammack, R. H. (1958). *Ann. Appl. Biol.* **46**, 186–197.
Carslaw, H. S. and Jaeger, J. C. (1959). "Conduction of Heat in Solids." Oxford Univ. Press (Clarendon), London and New York.
Chamberlain, A. C. (1967). *Q. J. R. Meteorol. Soc.* **93**, 509–515.
Chamberlain, A. C. (1975). *In* "Vegetation and the Atmosphere" (J. L. Monteith, ed.), Vol. 1, pp. 155–201. Academic Press, New York.
Chamberlain, A. C. and Chadwick, R. C., (1972). *Ann. Appl. Biol.* **71**, 141–158.
Champagne, F. H., Friehe, C. A., Larue, J. C., and Wyngaard, J. C. (1977). *J. Atmos. Sci.* **34**, 515–530.
Chock, D. P. and Dunker, A. M. (1983). *Atmos. Environ.* **17**, 11–24.
Cionco, R. M. (1972). *Boundary Layer Meteorol.* **3**, 255–263.
Cionco, R. M., Ohmstede, W. D. and Appleby, J. F. (1963). *Meteorol. Res. Notes No. 5.* USAERDAA, Fort Huachuca, Arizona.
Counihan, J. (1975). *Atmos. Environ.* **9**, 871–905.
Csanady, G. T. (1963). *J. Atmos, Sci.* **20**, 201–208.
Csanady, G. T. (1973). "Turbulent Diffusion in the Environment." Reidel, Dordrecht, The Netherlands.
Denmead, O. T. (1975). *In* "Vegetation and the Atmosphere" (J. L. Monteith, ed.), Vol. 2, pp. 1–31. Academic Press, New York.
Dollard, G. J. and Unsworth, M. H. (1983). *Atmos. Environ.* **17**, 775–780.
Edmonds, R. L. and Driver, C. H. (1974). *Phytopathology* **64**, 1313–1321.
Fisher, B. E. A. (1979). *In* "Mathematical Modelling of Turbulent Diffusion in the Environment" (C. J. Harris, ed.), pp. 237–275. Academic Press, New York.
Fitt, B. D. L. and Bainbridge, A. (1983). *Phytopathol. Z.* **106**, 214–225.
Fried, P. M., MacKenzie, D. R. and Nelson, R. R. (1979). *Phytopathol. Z.* **95**, 140–150.
Fuchs, N. A. (1964). "The Mechanics of Aerosols" (Transl. ed. C. N. Davis). Pergamon, Oxford.
Gifford, G. A. (1976). *Nucl. Saf.* **17**, 68–86.
Grace, J. (1977). "Plant Response to Wind." Academic Press, New York.

Graham, D. C., Quinn, C. E. and Bradley, L. F. (1977). *J. Appl. Bacteriol.* **43**, 413–424.
Gregory, P. H. (1945). *Trans. Br. Mycol. Soc.* **28**, 26–72.
Gregory, P. H. (1948). *Ann. Appl. Biol.* **35**, 412–417.
Gregory, P. H. (1968). *Annu. Rev. Phytopathol.* **6**, 189–212.
Gregory, P. H. (1973). "The Microbiology of the Atmosphere." Leonard Hill, London.
Gregory, P. H. (1982). *In* "Advancing Frontiers of Mycology and Plant Pathology" (K. S. Bilgrami, R. S. Misra, and P. C. Misra, eds.), pp. 107–117. Today and Tomorrow's Publ., New Delhi.
Gregory, P. H. and Lacey, M. E. (1964). *Trans. Br. Mycol. Soc.* **47**, 25–30.
Gregory, P. H. and Stedman, O. J. (1953). *Ann. Appl. Biol.* **40**, 651–674.
Gregory, P. H., Longhurst, T. J. and Sreeramulu, T. (1961). *Ann. Appl. Biol.* **49**, 645–658.
Gryning, S.-E., van Ulden, A. P. and Larsen, S. E. (1983). *Q. J. R. Meteorol. Soc.* **109**, 355–364.
Hall, C. D. (1975). *Q. J. R. Meteorol. Soc.* **101**, 235–244.
Hanna, S. R. (1982). *In* "Atmospheric Turbulence and Air Pollution Modelling" (F. T. M. Nieuwstadt and H. van Dop, eds.), pp. 275–310. Reidel, Dordrecht, The Netherlands.
Hanna, S. R., Briggs, G. A., Deardoff, J., Egan, B. A., Gifford, F. A. and Pasquill, F. (1977). *Bull. Am. Meteorol. Soc.* **58**, 1305–1309.
Harris, C. J. (ed.) (1979). "Mathematical Modelling of Turbulent Diffusion in the Environment." Academic Press, New York.
Hassid, S. (1983). *Atmos. Environ.* **17**, 523–527.
Hirst, J. M. and Stedman, O. J. (1963). *J. Gen. Microbiol.* **33**, 335–344.
Holtslag, A. A. M. and van Ulden, A. P. (1983). *J. Climate Appl. Meteorol.* **22**, 517–529.
Horst, T. W. (1977). *Atmos. Environ.* **11**, 41–46.
Hosker, R. P. and Lindberg, S. E. (1982). *Atmos. Environ.* **16**, 889–910.
Huang, C. H. (1979). *Atmos. Environ.* **13**, 453–463.
Ingold, C. T. (1971). "Fungal Spores; Their Liberation and Dispersal." Oxford Univ. Press (Clarendon), London and New York.
Itier, B. and Pauvert, P. (1979). *Bull. OEPP.* **9**, 251–264.
Jarvis, P. G., James, G. B. and Landsberg, J.-J. (1976). *In* "Vegetation and the Atmosphere" (J. L. Monteith, ed.), Vol. 2, pp. 171–240. Academic Press, New York.
Jeger, M. J. (1983). *Plant Pathol.* **32**, 5–11.
Jeger, M. J., Jones, D. G., and Griffiths, E. (1983). *Ann. Appl. Biol.* **102**, 237–244.
Jenkyn, J. F., Bainbridge, A., Dyke, G. V. and Todd, A. D. (1979). *Ann. Appl. Biol.* **92**, 11–28.
Johnson, A. G., and Dickson, J. G. (1919). *Wis. Agric. Exp. Stn. Bull.* **304**, 1–16.
Johnson, C. E., Biscoe, P. V., Clark, J. A. and Littleton, E. J. (1976). *Agric. Meteorol.* **16**, 17–35.
Johnson, K. B. and Powelson, M. L. (1983). *Phytopathology* **73**, 741–746.
Jones, A. M. (1983). Droplet deposition on elements of wheat plants. Ph.D. Thesis, University of Nottingham, Nottingham, England.
Kable, P. F., Fried, P. M. and MacKenzie, D. R. (1980). *Phytopathology* **70**, 601–604.
Kampmeijer, P. and Zadoks, J. C. (1977). "EPIMUL, a Simulator of Foci and Epidemics in Mixtures of Resistant and Susceptible Plants, Mosaics and Multilines." Pudoc, Wageningen.
Kiyosawa, S. and Shiyomi, M. (1972). *Nippon Shokubutsu Byori Gakkaiho* **38**, 41–51.
Lambert, D. H., Villareal, R. L. and MacKenzie, D. R. (1980). *Phytopathol. Z.* **98**, 150–154.
Lambert, E. B. (1929). *Phytopathology* **19**, 1–71.
Landsberg, J. J. and James, G. B. (1971). *J. Appl. Ecol.* **8**, 729–741.
Legg, B. J. (1975). *Q. J. R. Meteorol. Soc.* **101**, 597–610.
Legg, B. J. (1983a). *Philos. Trans. R. Soc. London Ser. B.* **302**, 559–574.
Legg, B. J. (1983b). *Q. J. R. Meteorol. Soc.* **109**, 645–660.
Legg, B. J. and Bainbridge, A. (1978). *In* "Plant Disease Epidemiology" (P. R. Scott and A. Bainbridge, eds.), pp. 103–110. Blackwell, Oxford.

Legg, B. J. and Long, I. F. (1975). *Q. J. R. Meteorol. Soc.* **101**, 611–628.
Legg, B. J. and Powell, F. A. (1979). *Agric. Meteorol.* **20**, 47–67.
Legg, B. J. and Price, R. I. (1980). *Atmos. Environ.* **14**, 305–309.
Legg, B. J. and Raupach, M. R. (1982). *Boundary Layer Meteorol.* **24**, 3–13.
Lemon, E. R. (1965). *In* "Plant Physiology," Vol. IV, Part A, pp. 203–227. Academic Press, New York.
Ley, A. J. (1982). *Atmos. Environ.* **16**, 2799–2808.
Limasset, P. (1939). *Ann. Epiphyt.* **5**, 21–39.
Llewellyn, R. P. (1983). *Atmos. Environ.* **17**, 249–256.
Long, P. E. and Pepper, D. W. (1976). *Symp. Atmos. Turbul. Diff. Air Qual. 3rd, 1976.*
McCartney, H. A. and Bainbridge, A. (1984). *Phytopathol. Z.* **109**, 219–236.
McCartney, H. A. and Fitt, B. D. L. (1985). *In* "Populations of Plant Pathogens: Their Dynamics and Genetics." (M. S. Wolfe and C. E. Caton, eds.), Blackwell, Oxford, in press.
McCartney, H. A., Bainbridge, A. and Aylor, D. E. (1983). *Bull. OEPP* **13**, 133–137.
McIntosh, D. H. and Thom, A. S. (1972). "Essentials of Meteorology." Wykeham Publ., London.
Maitani, T. (1977). *Boundary Layer Meteorol.* **12**, 405–423.
Maul, P. R. (1975). Central Electricity Generating Board Rep. SSD/MID/R28/75. U.K. Central Electricity Board, Leatherhead.
Maul, P. R. (1977). *Atmos. Environ.* **11**, 1191–1195.
May, K. R. and Clifford, R. (1967). *Ann. Occup. Hyg.* **10**, 83–95.
Mitchell, A. E. (1982). *Atmos. Environ.* **16**, 765–773.
Monsi, M., Uchijima, Z. and Oikawa, T. (1973). *Annu. Rev. Ecol. Syst.* **4**, 301–327.
Monteith, J. L. (1973). "Principles of Environmental Physics." Arnold, London.
Motha, R. P., Verma, S. B. and Rosenberg, J. (1980). *Agric. Meteorol.* **22**, 121–127.
Newhall, A. G. (1938). *Phytopathology* **28**, 257–269.
Nieuwstadt, F. T. M. (1980). *Atmos. Environ.* **14**, 1361–1364.
Nieuwstadt, F. T. M. and van Dop, H. (eds.) (1982). "Atmospheric Turbulence and Air Pollution Modelling." Reidel, Dordrecht, The Netherlands.
Oort, A. J. P. (1936). *Tijdschr. Plantenziekten* **42**, 179–234.
Oort, A. J. P. (1940). *Tijdschr. Plantenziekten* **46**, 1–18.
Panofsky, H. A. (1963). *Q. J. R. Meteorol. Soc.* **89**, 85–94.
Panofsky, H. A. (1973). *In* "Workshop on Micrometeorology." (D. A. Haugen, ed.), pp. 151–176. Am. Meteorol. Soc., Boston.
Pape, H. and Rademacher, B. (1934). *Angew. Bot.* **16**, 115–150.
Pasquill, F. (1962). "Atmospheric Diffusion," 1st Ed. Van Nostrand–Reinhold, Princeton, New Jersey.
Pasquill, F. (1974). "Atmospheric Diffusion," 2nd Ed. Ellis Harwood, Chichester.
Pasquill, F. (1976). Atmospheric dispersion parameters in Gaussian plume modelling. Part II. Possible requirements for change in the Turner Workbook Values U.S. Environ. Prot. Agency Publ. EPA-600/4-76.030B. Environ. Prot. Agency, Washington D.C.
Pasquill, F. and Smith, F. B. (1971). *Proc. Int. Clean Air Congr. 2nd, 1970*, pp. 1067–1072.
Paysour, R. E. and Fry, W. E. (1983). *Phytopathology* **73**, 1014–1020.
Perrier, E. R., Robertson, J. M., Millington, R. J., and Peters, D. B. (1972). *Agric. Meteorol.* **10**, 421–442.
Rauner, J. L. (1976). *In* "Vegetation and the Atmosphere" (J. L. Monteith, ed.), Vol. 2, pp. 241–264. Academic Press, New York.
Raupach, M. R. and Thom, A. S. (1981). *Annu. Rev. Fluid Mech.* **13**, 97–129.
Raynor, G. S., Hayes, J. V. and Ogden, E. C. (1970). Experimental data on dispersion and deposition of Timothy and corn pollen from known sources. Brookhaven Natl. Lab. Rep. BNL 50266 (T-595). Brookhaven Natl. Lab., Brookhaven, New York.

Raynor, G. S., Ogden, E. C. and Hayes, J. V. (1973). *Agric. Meteorol.* **11**, 177–195.
Reddi, S. C. (1975). *J. Palynol.* **11**, 111–120.
Reddi, S. C. (1976). *Proc. Indian Acad. Sci. Sect. B* **84**, 1–5.
Rowe, R. C. and Powelson, R. L. (1973). *Phytopathology* **63**, 984–988.
Schayes, G. (1982). *Atmos. Environ.* **16**, 1407–1413.
Schmidt, R. A., Carey, W. A. and Hollis, C. A. (1982). *Phytopathology* **72**, 1485–1489.
Scriven, R. A. and Fisher, B. E. A. (1975). *Atmos. Environ.* **9**, 49–58.
Sedefian, L. and Bennett, E. (1980). *Atmos. Environ.* **14**, 741–750.
Shaw, R. H. (1977). *J. Appl. Meteorol.* **16**, 514–521.
Shaw, R. H., Ward, D. P. and Aylor, D. E. (1979). *J. Appl. Meteorol.* **18**, 167–171.
Slinn, W. G. N. (1982). *Atmos. Environ.* **16**, 1785–1794.
Smith, F. B. (1959). *In* "Atmospheric Diffusion and Air Pollution—Advances in Geophysics" (F. N. Frenkiel and P. A. Sheppard, eds.), Vol. 6, pp. 193–210. Academic Press, New York.
Smith, F. B. and Blackall, R. M. (1979). *In* "Mathematical Modelling of Turbulent Diffusion in the Environment" (C. J. Harris, ed.), pp. 201–235. Academic Press, New York.
Smith, M. E. and Singer, I. A. (1966). *J. Appl. Meteorol.* **5**, 631–639.
Sreeramulu, T. and Ramalingam, A. (1961). *Ann. Appl. Biol.* **49**, 659–670.
Stedman, O. J. (1979). *Ann. Appl. Biol.* **91**, 271–285.
Stedman, O. J. (1980a). *Agric. Meteorol.* **21**, 111–127.
Stedman, O. J. (1980b). *Agric. Meteorol.* **21**, 195–203.
Stepanov, K. M. (1935). *Tr. Zashch. Rast. Ser. 2* **8**, 1–68.
Sutton, J. C., Gillespie, T. J. and Hildebrand, P. D. (1984). *Plant. Dis.* **68**, 78–84.
Sutton, O. G. (1953). "Micrometeorology." McGraw-Hill, New York.
Tang, C.-M. (1969). *Pure. Appl. Geophys.* **77**, 61–69.
Taylor, G. I. (1915). *Philos. Trans. R. Soc. London, Ser. A* **215**, 1–26.
Tennekes, H. and Lumley, J. L. (1972). "A First Course in Turbulence." MIT Press, Cambridge, Massachusetts.
Thom, A. S. (1971). *Q. J. R. Meteorol. Soc.* **97**, 414–428.
Thom, A. S. (1975). *In* "Vegetation and the Atmosphere" (J. L. Monteith, ed.), Vol. 1, pp. 57–109. Academic Press, New York.
Thompson, N. (1983). *Meteorol. Mag.* **112**, 249–260.
Thompson, N. and Ley, A. J. (1982). *Proc. Br. Crop Prot. Conf.—Weeds, 1982*, pp. 1039–1044.
Turner, D. B. (1970). Workbook of atmospheric dispersion estimates. U.S. Public Health Serv. Publ. 999 AP 26. U.S. Public Health Serv., Cincinnati.
Uchijima, Z. (1975). *In* "Vegetation and the Atmosphere" (J. L. Monteith, ed.), Vol. 2, pp. 33–64. Academic Press, New York.
Van der Hoven, I. (1968). *In* "Meteorology and Atomic Energy, 1968" (D. Slade, ed.). U.S. Atomic Comm., Publ. TID-24190. U.S. Atomic Comm., Washington, D.C.
van Ulden, A. P. (1978). *Atmos. Environ.* **12**, 2125–2129.
Waggoner, P. E. (1983). *Philos. Trans. R. Soc. London Ser. B* **302**, 451–462.
Wallace, E. R. (1934). *Rep. 1933 Bulb Res. Subcomm., Holland County Counc. (England)*, pp. 40–41.
Wilson, J. D., Thurtell, G. W. and Kidd, G. E. (1981). *Boundary Layer Meteorol.* **21**, 423–441.
Wippermann, F. (1972). *Contrib. Atmos. Phys.* **45**, 148–163.
Wyngaard, J. C. (1973). *In* "Workshop in Micrometeorology" (D. A. Haugen, ed.), pp. 101–149. Am. Meteorol. Soc., Boston.
Yaglom, A. M. (1977). *Boundary Layer Meteorol.* **11**, 89–102.
Yokahama, O., Gamo, M. and Yamamoto, S. (1977). *J. Meteorol. Soc. Jpn.* **55**, 312–318.

6

GENETIC MODELLING FOR CROP DISEASE MANAGEMENT

D. R. MacKenzie* and M. H. Royer†

*Department of Plant Pathology and Crop Physiology, Louisiana State University, Baton Rouge, Louisiana, USA
†Plant Disease Research Laboratory, USDA-ARS, Frederick, Maryland, USA

I. INTRODUCTION

It has long been the dream of crop protection specialists to develop pest-resistant cultivars that could be substituted for existing susceptible cultivars. For many plant disease problems this mission remains a dream, but patterns of success and failure are now forming. These present a view of what is happening genetically, and allow us to project our expectations to other problems. Such projections come in the form of models, both conceptual and mathematical, which permit the extension of existing knowledge into the relatively unknown. Modelling the genetics of plant disease relationships has taken on a growing importance as better ways are sought to protect crops from destructive parasites. The intention of this chapter is to describe the genetic problems arising from the use of host plant resistance and the current 'state of the art' in genetic models for crop disease management.

There are many examples of the successful use of host plant resistance in agriculture (Eenink, 1976). Often successes are overshadowed by the frustrating search for solutions to long-term problems. As information accumulates on major pests of crops, it has become apparent that the principles underlying host plant resistance and pathogen variation extend far beyond plant pathology to examples of insects, nematodes and other pests of crops. In this chapter the specific examples drawn from the plant pathological literature will be chosen for their applicability to crop parasites in general.

Breeding for resistance in crops is historically interesting and quite recent on the time scale of agriculture. Although very little is known about the evolution of cultivars, most would agree that successful plantings of a crop were perpetuated by farmers and disseminated to neighbours in a long and continuing process. In areas subject to pest problems, part of the success of a cultivar would be its performance under the growing conditions of that particular area. From this process the 'land races' of crops came to be established. Many of these are thought to possess pest resistance as well as many other desirable attributes.

With the advent of the application of the scientific method to agriculture, the development and maintenance of crop cultivars took on new meaning. During the early to mid-1800s, specialists began to extract systematically and stabilise genetically plant characteristics that were of special interest to farmers and/or consumers. By the turn of the century, increasing recognition was given to progeny differences in response to various types of stresses. With the re-discovery of Mendel's laws at this time, it was inevitable that genetic studies of progeny differences would uncover the fact that single genes in the host were very effective in controlling infection by some fungi (Johnson, 1961). Great effort was then devoted to incorporating single genes for resistance into crops for the control of some major pest problems.

In many cases single gene resistance to plant pathogens was very short lived. The resistance was said to have 'broken', and soon some scientists began to speak of the futility of single gene resistance. The 'breakdown' of resistance was due to genetic variation in the pathogen population. Flor (1955), working with rust of flax (caused by *Melampsora lini*), observed a gene-for-gene relationship which has now been extended as a universal principle for host–parasite interactions including fungi, nematodes and some insects. Flor discovered that, for each gene for specific resistance in the host, there may be a specific gene for virulence in the parasite which can overcome that resistance. The pattern appears so universal in agricultural crops that some suggest that genetic analyses of parasite populations can be determined indirectly by knowing the genetics of the host (Wolfe *et al.*, 1976). Others have extended the gene-for-gene theory to suggest that polygenic host resistance (determined by many genes) can only be overcome by parasitic populations with complementary genes for added virulence. This extension of Flor's gene-for-gene concept awaits scientific proof, but it has become a key assumption of many breeding programmes striving for polygenic resistance to parasites.

The source of variation in the parasite population is said to be mutation. The magnitude of the mutation rate for these events is most often only speculative, as the measurement of mutation rates in lower organisms is very difficult. The observed frequency of mutant types in culture is only an indication of the mutation rate, since the measurement of mitotic generations for non-synchronous nuclear divisions is almost impossible. Without this information, a distinction cannot be made between two mutation events and one event followed by mitotic division. Hence, most descriptions of mutations in lower organisms speak of mutational frequencies as a measure of the expected occurrence of new types in a given population.

Much of the information on mutational frequencies has come from limited laboratory studies. The application of this information to the understanding of the host–parasite interactions is open to question. Many propose mutational frequencies of around one in a million. This would provide ample opportunity for variation in the parasite population as the sheer numbers of these organisms, even in one field, are sometimes substantial. Vanderplank (1982) reviews calculations which support this view.

More complex forms of a parasite population can be produced through sexual recombination, which reassorts genes through segregation. This was a primary consideration in the North American barberry eradication programme for wheat stem rust control (Roelfs, 1982), an effort which attempted to reduce the stem rust causing genetic variation in the *Puccinia graminis tritici* population by removing the alternate host, barberry. Subsequently, it was discovered that the population was provided with another opportunity to produce variation through the parasexual cycle (see Burnett, 1975).

Relatively little genetic study has been made of gene linkages and gene recombination frequencies for parasites (Van der Plank, 1968; McKey, 1973; Burnett, 1975; Eenink, 1976; Leonard, 1977) and much of the genetic theory for plant pathogens has been extended from microbiological research.

The reassortment of genes giving new combinations leading to parasitic variation was seen by early researchers, who began describing 'physiological' races. Now more appropriately called pathological races, these combinations are seen to exist in very large numbers for several pathogens and insect pests. The explanation for this apparent extensive variation comes from the unique system of nomenclature used for pathological races. This system identifies all possible genetic combinations as unique variants. For a pathogen possessing 10 genes for virulence, one would expect that 2^{10} or 1024 races may exist; for 40 genes, one would expect billions of races. This unique system of taxonomy soon becomes clumsy and contributes to the frustration of trying to deal with highly variable pathogens and in the search for stable host plant resistance (Vanderplank, 1982).

Biological fitness (w) deals with the generation-to-generation reproductive success of a genotype including disease/pest cycles, overwintering and mating. It embraces all aspects of why some genotypes predominate over others, but it is a collective expression that often lacks explanation or predictive value. For this reason, most current modelling efforts are aimed at understanding parasitic fitness (w'), which focuses only on an organism's competitive ability during pathogenesis and disregards other factors such as overwintering during saprophytic stages. The common experimental technique to estimate (w') is to mix two isolates of a parasite and, over generations, test for shifts in the proportion of the two types. It is usual to express the fitness of the less fit isolate relative to the more fit by giving a value of (w') between 1 (of equal fitness) and 0 (no fitness). The coefficient of selection (s) is defined as $(1 - w')$ and is very useful in constructing genetic models of the dynamics of the host–parasite interaction (for more detail see MacKenzie, 1978, 1980).

Two approaches have been used to try to address the question of how to improve efficiency in the deployment of genetic resistance to parasites. The first approach attempts to understand, through modelling, the processes of co-evolution between a wild host and a parasite. The second looks to the agricultural production systems for models of why some efforts have been successful and others have not.

II. LESSONS FROM NATURE

A. Wild Populations

A generally accepted attribute of wild populations is their diversity (Browning and Frey, 1969; Robinson, 1980). Mixtures of plant species and genetic heterogeneity within a species are both used to explain the stability of natural systems in response to all types of stresses. This characteristic is seen as an important element by those who have developed models to explain the success of wild populations. Such models have been used to develop proposals for the deployment of genes for resistance in agricultural situations.

Several mathematical models have been developed to express the relationship between a wild host population and a parasite population and the consequences of single gene resistance. In the simplest form, it is argued that as an allele for resistance increases in a host population, selection pressure will act to increase the frequency of alleles for virulence in the parasite population. Individuals carrying the virulence allele will, of course, reproduce at a greater rate than individuals with the avirulence allele. Subsequently, the increased frequency of individuals in the parasite population able to infect and reproduce on the once-resistant host genotype will act to select against that host genotype. This will then favour other genetic combinations in the host population that possess resistance to the now predominant pathogen genotypes. Thus an ebb and flow of host and pathogen genotypes will continue as changing frequencies respond to selection in both populations.

The existence of various genotypes as a mixture in a population is called polymorphism. Many genetic modellers now subscribe to the view that it is desirable to use polymorphism to stabilise the ebb and flow of genetic types within that population. Much effort has, therefore, been devoted to understanding polymorphism. Two types of polymorphism are recognised, and this division is essential to modelling the genetics of crop disease management.

Transient polymorphism results from the ebb and flow of selection that takes place when specific genotypes of the parasite overcome host plant resistance and increase in number. It also occurs when specific host plant resistance successfully blocks specific genotypes of the parasite and increases in numbers in the plant population. This form of selection, directional selection, can result in so-called 'boom and bust cycles' in agriculture and is obviously very undesirable.

Balanced polymorphism results when the selection pressures acting to increase one genotype at the expense of another are balanced by selection pressures acting against that genotype. This has obvious applications to agricultural production in that, if a balance can be established to maintain certain races at low frequencies, a parasite population could be managed, through crop management strategies, to acceptable levels.

The contributions of transient and balanced polymorphism to the stability and/or to the success of plants in wild populations is not known. Far too few studies have been conducted on wild species. Robinson (1980) refers to this as a 'major gap in scientific knowledge'. It is this absence of knowledge that has caused so many to attempt to model the host–parasite interaction mathematically to search for methods for managing parasite populations.

One of the first theoretical attempts to analyse genetic interactions between host and parasite was by Mode (1958). An excellent summary of his contributions was presented by Leonard and Czochor (1980) and is condensed here from that review article. Mode developed models that explored the conditions under which a balanced or stable polymorphism could be maintained. In this first model Mode (1958) explored the consequences of two genes for resistance in the host and two corresponding genes for virulence in the parasite. The genes for resistance were allelic at a single locus. The genes for virulence were assumed to occur at independent loci, where they were distinguished from alleles for avirulence. Mode assumed that the fitness of the host would vary inversely with the fitness of the parasite. Moreover, in the genetically mixed populations of the parasite, a weighted fitness could be calculated from the frequencies of the host genotypes for each of the parasite genotypes. Mode used differential equations to model the change in gene frequency. From his analyses he concluded that multiple alleles for resistance in the host could provide a stable equilibrium. Thus, plants heterozygous for two alleles for resistance fared better than genotypes homozygous for either allele. Mode's model, however, could not explain the persistance of genes for susceptibility in balanced polymorphic systems and the model was only applicable to open-pollinated host species.

Mode (1960) overcame the restrictions of the host pollinating system in a second model that showed how the fitness of the host cultivars determined the frequencies of parasite races and how the fitness of parasite races determined the frequency of host cultivars at equilibrium. Subsequent models by Jayakar (1970) and Leonard (1977) supported these relationships.

B. Multiline Varieties

The models discussed previously for gene-for-gene interaction imply that the selection of new races of the parasite would be slowed by the use of genetically diverse crop plantings as mixtures of cultivars or near isogenic lines. Mixtures of near isogenic lines are called multiline varieties and were first proposed by Jensen (1952) and Borlaug (1953). Through multilines, they argued, diversity of wild populations could be mimicked in agricultural production. By mixing together phenotypically similar lines known to differ in their genetic resistance to the target parasite populations, stable resistance could be attained (for a review of the concept see Browning and Frey, 1969). Skeptics have argued that multiline varieties will simply provide a substrate for the selection of 'superraces' of the parasite which can grow and reproduce on all components of the mixture. Just how quickly superraces would evolve, if ever, is an important question concerning the use of multilines in agriculture.

Two schools of thought exist on how multilines ought to be deployed. The first, in the terminology of Marshall (1977), is called 'dirty crop multilines'. By this reasoning, if some proportion of the multiline is intentionally susceptible to the genetically simple races of the parasite, selection of superraces will be slowed down and the life of the multiline thereby extended. These forces have been termed stabilising selection by Van der Plank (1968). Genetically complex races are thought, by this reasoning, to be relatively less fit than simple races and 'dirty crop multilines' would capture this benefit as a disease management strategy.

Proponents of the second school of thought, 'clean crop multilines', argue that the presence of a susceptible component will only serve as a substrate for the build-up of disease and for mutation and recombination. 'Clean croppers' see a need for constant replacement of components in a never-ending flux of pathogen shifts. This, of course, would greatly restrict multiline deployment in many agricultural uses, especially in countries with developing economies and limited seed production industries.

Considerable mathematical modelling has been devoted to trying to understand the genetic relationships of multilines to populations of parasites and to determine whether multilines have practical applications in agriculture. A model proposed by Groth (1976) well demonstrates the process of mathematically exploring the impact of multiline varieties on parasite selection. Groth conceived a multiline of n components mixed at equal frequency, with each component possessing a single gene for resistance to a pathogen. Various members of the pathogen population would be capable of overcoming single resistance genes with a corresponding gene for virulence. Without that corresponding gene, the pathogen would be unable to reproduce. But, with each additional gene, the fitness of that genotype would be reduced by the same coefficient of selection s. Hence, for every additional gene in a parasite's genotype, it could attack a greater number of components in the multiline variety, but only at a cost of reduced fitness. Thus the most successful races would be represented by those genotypes which optimise some number of genes for virulence by not sacrificing too much fitness with 'unnecessary genes for virulence'. Groth concluded that the balance would be found in those parasite genotypes for which the number of virulence genes (m) would be calculated as $(1 + s)/2s$ when the losses in fitness for additional genes for virulence were additive. When those losses were multiplicative, Groth reasoned that maximum fitness would be at $m = 1/\ln(1 - s)$. Hence, these relationships seemed independent of the number of components in the multiline and determined only by the coefficient of selection (s) acting on the pathogen population.

Groth's model was elaborated on by Barrett and Wolfe (1978), who allowed for changes in the coefficient of selection over generations. With this modification, the races of greatest fitness were determined not only by the coefficient of selection, but also by the number of components in that multiline. Although solving some problems, the new model introduces others and it is still not clear if multiline varieties will 'stabilise' a parasite population and result in a desired level of balanced polymorphism.

Another desirable attribute of multiline varieties is said to be that they delay the rate of disease build-up in the mixed population. Borlaug (1959) likened a multiline to green grass in a prairie fire, slowing the progress of an epidemic. Mathematical models of this relationship have been attempted, with the first success by Leonard (1969), who modelled multiline varieties in an effort to describe the effects of a multiline on a epidemic's rate of disease increase. From field-plot data, Leonard demonstrated that the rate of increase of stem rust of oats, caused by *Puccinia graminis avenae*, in a multiline (r_m) was equal to the rate of increase in an entirely susceptible plot (r_s) plus the reciprocal of the pathogen's generation time in days (c) times the natural log of the proportion of susceptible plants in the multiline (m). This model was written as

$$r_m = r_s + c[\ln(m)]$$

This equation shows the diminishing return on slowing an epidemic for each additional component line included in a multiline. Moreover, the genetic and economic implications are that at some point the benefits of an additional component line (or adding one more gene) in a multiline will not justify the risk of exposing the resistance gene to the pathogen population or the additional expense incurred in breeding and maintaining the more complex multiline.

Verification of Leonard's model was needed to see if it was more generally applicable. Elliott (1981), working with the causal agent of wheat powdery mildew, *Erysiphe graminis tritici*, showed that the return in benefits from additional component lines, up to a maximum of 50 ($m = 0.02$), did indeed follow Leonard's model.

Multilines alone seem inadequate to the task of managing many foliar parasites, as the number of genes available is often limited. Integrated strategies have, therefore, been explored in an attempt to complement the benefits of multilines in order to manage target parasites at acceptable levels. Computer simulations of these complex models have been very helpful in looking at complex options. Pseudo-three-dimensional graphic displays have to be used to visualise the consequences of various strategies on managing a foliar pathogen of small grains (MacKenzie, 1979). The following section looks at how one model was developed as an example of how the task can be approached.

The standard epidemiological model for disease increase (Van der Plank, 1963) provides for the exponential increase of the proportion of diseased tissue (x) over time (t) and corrects for the unavailability of tissue as the epidemic progresses (multiple infection). It would be written as

$$x/(1 - x) = [x_0/(1 - x_0)] \exp(rt)$$

where x_0 is the initial inoculum, r is the apparent infection rate, and 1 is the total amount of tissue, healthy or otherwise. The model does not take into account disease spread as a function of distance. That is, every potential site within a field is said to have an equal chance of becoming infected. Obviously, this is not the case. In most disease situations, closer proximity to the source of inoculum tends to increase the probability of receiving inoculum, which increases the probability of that site becoming diseased. This relationship can be expressed as

$$x = a/D^b$$

which says that disease (x) will vary inversely as some power (b) of distance (D) from a point source of strength (a). Merging this model with Van der Plank's (1963) is an algebraic nightmare (MacKenzie, 1979). It does, however, demonstrate the application of computer modelling in understanding gene deployment strategies.

Using a mainframe computer, one can electronically describe a large field of a crop planted as a multiline variety and then study the consequences when 'planted' in patterns or at random. Realistic numbers of potential infection sites on each 'plant' can be described and 'inoculum' can be introduced into the system at various points in the 'field'. By knowing the geometric relationship of each plant to all others, the expected rate of increase on each plant due to reinfection of the same plant, the genetic compatibility of the parasite races on each plant within the field, and the probability of spread between plants, epidemics can be simulated and the resulting amount of disease tabulated.

Such a simulation has been done for foliar diseases on small grains, with some interesting results (MacKenzie, 1979). Firstly, the effects of the number of components in a multiline are directly associated with reduced infection rates as predicted by Leonard (1969) and verified by Elliott (1981). Secondly, the relationship between the number of component lines on the apparent infection rate (r) is not linear over the course of the epidemic. This complex relationship is not understood. It does, however, point out the difficulties of anyone presently drawing hard conclusions on the benefits of multiline varieties.

Although computer simulation has received a lot of criticism within the scientific community, one often overlooked benefit is the use of the method in designing appropriate experiments to answer questions of genetic relevance. Computer simulations of multilines served as the source of questions for studies by Elliott (1981). He then investigated those questions by field experimentations on multilines. The importance of the exercise was not that the simulations proved anything. They simply provided one quick way of addressing complex relations in search of the consequences, likely outcomes, and appropriate field plot design to investigate those relationships.

Some theories propose that the component lines in a multiline should possess more than one gene to stabilise the pathogen population, and thereby suppress the potential for superrace development and, perhaps, better manage an epidemic's development. After all, it is observed, wild populations are rarely structured with one gene for resistance per plant. Computer modelling of these multiline options is one of the most reasonable ways of trying to understand their use in agriculture. This is because field research on so many genetic possibilities seems out of the question.

C. Gene Pyramids

As an alternative to multiline varieties, some individuals have proposed that genes be 'heaped together' into single cultivars for better protection. The argument is based on a model which addresses the relative ease with which a pathogen might attain superrace status. On a multiline variety, single event mutations can contribute additively to the evolution of a superrace. On a six-component multiline, six individual mutations would be required to develop a superrace. For an assumed individual gene mutation frequency of 1×10^{-6}, the probability that any one of six alleles might arise would be the sum of the probabilities of the individual events, or 6×10^{-6}. Moreover, dirty crop multilines might only aggravate this situation by providing a susceptible substrate for the multiplication of the pathogen and hence greater chance for change.

For a six-gene pyramid, six simultaneous mutations would be required to overcome that resistance. The argument continues that the frequency of this rare event will be 1×10^{-36}, or the product of the individual gene mutation frequencies. Hence, it is argued from this model that gene pyramids would be far superior to multiline varieties for stalling superrace evolution.

A second argument in support of gene pyramids was the discovery, first by Martin and Ellingboe (1976) and then extended by Nass *et al.* (1981), who reported on the residual effects of genes for resistance that have been successfully challenged by a compatible race of the pathogen. Martin and Ellingboe (1976) discovered that when challenged with compatible races of *E. graminis tritici*, wheat lines possessing a 'broken' resistance gene were less susceptible than were susceptible lines without the gene. Using near-isogenic lines of 'Chancellor' wheat, their research demonstrated that pathogens overcoming the resistance of a host do not return that relationship to 'square zero' as originally assumed. Importantly, a slight benefit for the host is retained. When these slight differences in reduced sporulation, extended latent period and lessened disease efficiency are added together, a significant reduction in the rate of epidemic development is expected. Genetic modelling of these gene pyramids is under way to see if this might be one way to manage epidemics even with the existence of superraces. But complications have been discovered that make simple modelling seem unrealistic. From the results of Nass *et al.* (1981) it is evident that no two host genes for resistance act identically after they have been 'defeated' by the pathogen. Further work by Royer *et al.* (1984a) demonstrated a significant near-isogenic line by isolate interaction. This complicates even more any effort to model the interactions between the host and parasite. This leads to the question, 'can general models be made from the wild host–parasite interaction that will allow us to deploy crop cultivars in patterns mimicking wild species?' At this time it is difficult to decide.

III. LESSONS FROM AGRICULTURE

A. Agricultural Surveys

Some specific criticism has been directed at the modelling of wild populations for applications in agricultural production. The brunt of that criticism has been directed at the obvious differences between wild populations and agricultural plantings. It is said that the physically isolated, poorly nurtured and generally less thrifty wild plant may be successful in a wild setting but, under the stresses of intensive agricultural production, with high soil fertility, ample moisture and physical crowding, some of the best wild genotypes may be unacceptably susceptible to many parasites. Given this perspective, many individuals have attempted to look at the reasons why some cultivars' genetic resistance has endured for long periods while other genetic combinations have produced failure. As an example of this approach to modelling agricultural genetic systems, we will use the story of 'Selkirk' spring wheat.

'Selkirk' was grown extensively in Canada for 18 years with little or no stem rust disease. During this period, extensive surveys for rust of wheat in North America were carried out by the United States Department of Agriculture's Rust Laboratory in St. Paul, Minnesota and by the Canadian Department of Agriculture's Rust Laboratory in Winnipeg, Canada, Vanderplank (1975, 1982) used these yearly surveys to analyse the dynamics of the parasite population in the *Puccinia* pathway.

The cultivar 'Selkirk' is known to possess specific genes for resistance to stem rust of wheat which are identified as *Sr*6 and *Sr*9*d*, in addition to others. For a genotype of the pathogen to overcome the resistance of 'Selkirk', it must possess the corresponding combinations of genes for virulence. Isolates with this virulence combination are designated as 15B–3 and have occurred frequently in the mid-western United States. In Canada, this combination was very rare. Vanderplank (1975, 1982) concluded that isolates of the parasite carrying genes for virulence corresponding to *Sr*6 and *Sr*9*d* were less fit to survive in Canada. This he attributed to the forces of stabilising selection. Stabilising selection, he said, operates against the genetically complex races of the pathogen because they are less fit to survive (see section II,B). It was argued that, since race 15B–3 was commonly found on winter wheat varieties in the United States during this period, therc must be some selection pressure suppressing race 15B–3 in Canada.

A second explanation could be that the survey methods used in Canada are very different from those used in the United States. It is possible that the time of year during which the sampling is performed alters the ratio of races collected significantly. Moreover, the influence of host genetics on the sampled population may determine the mix of isolates collected. The predominance of collections in the United States would probably come from winter wheat cultivars, while those in Canada would almost exclusively come from spring wheat cultivars. Also, the frequency of sexual reproduction, if any, within a rust population may have some role in determining racial frequencies (Roelfs and Groth, 1980).

Finally, there exists the possibility that the sampling techniques employed by both organisations are inadequate to draw reasonable conclusions about shifts in the pathogen population. Although most annual surveys represented hundreds of isolates, statistical arguments can be offered that even that effort is inadequate to represent the variation that may exist in the *Puccinia* pathway. Research by Royer *et al.* (1984a) with *E. graminis tritici* collected over a 4-year period in Pennsylvania supports this point. Analyses of the survey information indicated that a far larger number of samples than their several hundred would be required to address specific hypotheses on the mixture of races in Pennsylvania, a relatively small area when contrasted with the *Puccinia* pathway.

G. J. Green (personal communication, August, 1980) discounted Vanderplank's stabilising selection explanation for the success of 'Selkirk'. He argued that the greenhouse temperatures in the St. Paul tests were too high during the early years of the United States survey data. Gene *Sr*6 is known to be temperature sensitive, which could account for the high frequency of race 15B–3 in the United States survey. The high greenhouse temperature problem has since been corrected.

A. P. Roelfs (personal communication, March, 1983) observed that the Canadian race survey erred, until recently, by not using an appropriate cultivar to detect virulence to *Sr9d*. The differential cultivar 'H44' was used instead of the United States standard differential 'Hope'. This has since been corrected, but the example points out the difficulty in determining from published survey data the response of a parasite population to selection pressures, particularly when the data are from different national survey programmes using different methods and protocols.

Another explanation for the success of 'Selkirk' may be that the resistance comes from the combined effects of *Sr*6 and *Sr9d* acting together to retard disease build-up (see section II,C). This would have the effect of reducing the proportion of rust isolates collected on 'Selkirk' during the yearly Canadian surveys.

A. P. Roelfs (personal communication, March, 1983) offers an alternative explanation. He has measured the attribute of 'adult plant' resistance to stem rust in 'Selkirk'. This form of resistance is expressed as a reduction in the number of lesions for a given number of spores applied and as fewer sites available for infection. This characteristic is expressed more intensely as the plant matures, giving the adult plants a degree of resistance not expressed in the seedling stage. Roelfs attributes this resistance to another gene, *Sr*2, and dismisses the stabilising selection explanation of Vanderplank (1975, 1982).

In summary, the success of 'Selkirk' in Canada remains an enigma, and the various models that have been applied to explain the existing information on host genetics and race surveys have not answered all the questions to everyone's satisfaction.

Surveys of late blight of potatoes, caused by *Phytophthora infestans*, in both North America and Europe have led to inconsistent results that have also baffled researchers but are easily explained by the etiology of potato late blight. *P. infestans*, in contrast to *P. graminis*, is not thought to spread over very long distances within a season. It moves predominantly as an infection, incipient and otherwise, of seed tubers. Any mutations or recombinations in the parasite population will probably be in close association with that particular planting. Surveys do little more than mirror the mutation and recombination frequencies occurring in the parasite population and the history of that seed lot. Hence, few conclusions can be drawn from the confusion found for national surveys of *P. infestans*.

The rice blast disease pathogen, *Pyricularia oryzae*, is similar to *P. infestans* in its 'plasticity' for genetic variation. Argument continues in the scientific community as to the source of this genetic plasticity. Until more is known about the genetics of *P. oryzae*, and in particular its ploidy level, national surveys on the population structure of *P. oryzae* cannot be interpreted.

These selected examples are offered to point out the need for much more information on the parasite, the host and the epidemiology of disease. Internationally acceptable protocols on race identification are needed if such diseases are ever to be adequately surveyed and modelled with the intention of understanding why some crop cultivars have stable resistance and how crop and pest management systems can be improved.

B. Land Races

The usual method of breeding for disease resistance is to hybridise an agronomically superior parent with a second parent possessing disease resistance, and then to select superior progeny through the segregating generations. Some scientists now question this procedure as the selected progeny rarely, if every, match the resistance of the original source. Perhaps it is a romantic notion, but the 'good' resistance of the old land races seems somehow superior. Efforts are now underway to examine in detail the old varieties and their attributes to pest resistance.

Information on the effects of crop management intensity on pest development will be an important contribution to understanding the durability of land races. Can old land races of rice withstand today's high soil fertility and resist blast disease? What are the trade-offs for more intensive production vs less disease? Genetic modelling of these questions will permit an evaluation at an earlier stage of cultivar development and perhaps point to ways to improve conventional selection procedures for breeding cultivars resistant to parasites.

One novel approach has been to induce single gene semidwarf mutations in old land races of rice and to study their performance under increasing levels of management relative to the tall parent. Evaluation of these results will be useful in judging the performance and usefulness of the old crop varieties in crop management programmes.

C. Rate-Limiting Resistance

Much of the previous discussion has concentrated on the qualitative responses of cultivars to disease. Very little is known about the genetics of quantitative traits such as rate-limiting resistance or the capacity of a parasite to overcome this form of resistance. Some authors have chosen to define rate-limiting resistance (RLR) as acting against all isolates of the pathogen. By this definition it is impossible for RLR to fail, because if it does, it is not RLR. Such tricks with the definitions, however, fail to address the real questions of whether RLR should be used and if it is durable.

Although little is known about the inheritance of RLR, considerable effort has been given to identifying its components (Johnson and Taylor, 1976; Parlevliet, 1979), which can be thought of as reduced sporulation, extended latent period and lessened disease efficiency. Our present understanding is that these components of RLR are under polygenic control, although exceptions exist. Cultivars which have RLR typically express it as less sporulation, although other components of RLR often contribute. Little genetic modelling of these traits has been attempted.

It seems reasonable to assume that the potential exists within a parasite population to overcome one or more of the components of RLR. Some evidence to support this view has been offered for wheat powdery mildew (Rouse *et al.*, 1980; Royer *et al.*, 1984b) and rice blast (Villareal *et al.*, 1981).

The speed with which rate-limiting resistance might 'erode' has been simulated through computer modelling for the range of differences in the host and in the parasite population. Construction of such models suggests that the speed of erosion of rate limiting resistance is of concern in managing crop varieties, but will not occur nearly as quickly as with the 'boom and bust cycle' of single gene resistance.

IV. CONCLUSIONS

Genetic modelling of host plant resistance and parasite variation has been very useful in addressing opportunities for research on how best to deploy genetic resources in agriculture. It is hoped that this chapter will encourage others to add to the search for better methods to protect agricultural crops from parasites.

REFERENCES

Barrett, J. A. and Wolfe, M. S. (1978). *Phytopathology* **68**, 1535–1537.

Borlaug, N. E. (1953). *Phytopathology* **43**, 467.

Borlaug, N. E. (1959). *Proc. Int. Symp. Wheat Genet., 1st*, pp. 12–26.

Browning, J. A. and Frey, K. J. (1969). *Annu. Rev. Phytopathol.* **7**, 355–382.

Burnett, J. H. (1975). 'Mycogenetics: An Introduction to the General Genetics of Fungi.' Wiley, New York.

Eenink, A. H. (1976). *Neth. J. Plant Pathol.* **82**, 133–154.

Elliott, V. J. (1981). The influence of the number of component lines and of rate-limiting resistance on the epidemiology of multiline cultivars. Ph.D. thesis, Department of Plant Pathology, The Pennsylvania State University, University Park.

Flor, H. H. (1955). *Phytopathology* **45**, 680–685.

Groth, J. V. (1976). *Phytopathology* **66**, 937–939.

Jayakar, S. C. (1970). *Theor. Popul. Biol.* **1**, 140–164.

Jensen, N. F. (1952). *Agron J.* **44**, 30–34.

Johnson, R. and Taylor, A. J. (1976). *Annu. Rev. Phytopathol.* **14**, 97–119.

Johnson, T. (1961). *Science* **133**, 257–362.

Leonard, K. J. (1969). *Phytopathology* **59**, 1845–1850.

Leonard, K. J. (1977). *Ann. N.Y. Acad. Sci.* **287**, 207–222.

Leonard, K. J. and Czochor, R. J. (1980). *Annu. Rev. Phytopathol.* **18**, 237–258.

MacKenzie, D. R. (1978). *Phytopathology* **68**, 9–13.

MacKenzie, D. R. (1979). *In* 'Proceedings of the Rice Blast Workshop.' International Rice Research Institute, Los Banos, The Philippines.

MacKenzie, D. R. (1980). *In* 'Breeding Plants Resistant to Insects' (F. G. Maxwell and P. R. Jennings, eds.), pp. 183–213. Wiley, New York.

McKey, J. (1973). *Proc. Int. Congr. Genet., 13th. Genetics* **74** (June Suppl., No. 2., Part 2), 169.

Marshall, D. R. (1977). *Ann. N.Y. Acad. Sci.* **287**, 1–20.

Martin, T. J. and Ellingboe, A. H. (1976). *Phytopathology* **66**, 1435–1438.

Mode, C. J. (1958). *Evolution* **12**, 158–165.

Mode, C. J. (1960). *In* 'Biometrical Genetics' (O. Kempthorne, ed.), pp. 84–97. Pergamon, Oxford.

Nass, H. A., Pederson, W. L., MacKenzie, D. R. and Nelson, R. R. (1981). *Phytopathology* **71**, 1315–1318.

Perlevliet, J. E. (1979). *Annu. Rev. Phytopathol.* **17**, 203–222.

Robinson, R. A. (1980). *Annu. Rev. Phytopathol.* **18**, 189–210.

Roelfs, A. P. (1982). *Plant Dis.* **66**, 177–181.

Roelfs, A. P. and Groth, J. V. (1980). *Phytopathology* **70**, 855–862.

Rouse, D. I., Nelson, R. R., MacKenzie, D. R. and Armitage, C. R. (1980). *Phytopathology* **70**, 1097–1100.

Royer, M. H., Nelson, R. R. and MacKenzie, D. R. (1984a). *Phytopathology* **74**, 1176–1180.

Royer, M. H., Nelson, R. R., MacKenzie, D. R. and Diehle, D. A. (1984b). *Phytopathology* **74**, 1001–1006.

Van der Plank, J. E. (1963). 'Plant Diseases: Epidemics and Control.' Academic Press, New York.

Van der Plank, J. E. (1968). 'Disease Resistance in Plants.' Academic Press, New York.

Vanderplank, J. E. (1975). 'Principles of Plant Infection.' Academic Press, New York.

Vanderplank, J. E. (1982). 'Host–Pathogen Interactions in Plant Disease.' Academic Press, New York.

Villareal, R. L., Nelson, R. R., MacKenzie, D. R. and Coffman, W. R. (1981). *Phytopathology* **71**, 608–611.

Wolfe, M. S., Barrett, J. A., Shattock, R. C., Shaw, D. S. and Whitbread R. (1976). *Ann. Appl. Biol.* **82**, 369–374.

7

CONSTRUCTION OF PREDICTIVE MODELS: I. FORECASTING DISEASE DEVELOPMENT

W. E. Fry* and G. R. Fohner†

*Department of Plant Pathology
†Department of Agricultural Economics
Cornell University, Ithaca, New York, USA

I. INTRODUCTION

The application of mathematical modelling is a recent development in the well-established effort to predict the need for disease management technology. It has only been since the mid-1970s that mathematical models contributed to specific, practical forecast efforts. Prior to that time mathematical models were used to evaluate general questions in plant pathology and to identify important new questions.

Achievement of the potential benefit of mathematical modelling to disease forecasts rests on knowledge of several factors: the objectives of forecasts, the epidemiological principles of disease forecasting and successful examples of forecasts. We therefore review these factors briefly before considering the contributions of mathematical models.

II. DISEASE FORECASTS: OBJECTIVES, PRINCIPLES AND EXAMPLES

Disease forecasts are predictions about the probable occurrence or non-occurrence of plant disease in economically important levels. The predictions are frequently associated with recommendations for disease suppression. The development of forecasts is motivated by several objectives. One objective is to increase farm income by providing more efficient allocation of disease management resources. Much effort has focused on one class of these resources, pesticides. A second objective for forecasts is to decrease risk of large losses in crop value from disease. Growers that are risk-averse use forecasts to indicate when additional resources may be needed to prevent large losses even if the use of those resources based on forecasts reduces average net income. A third objective is to decrease the amount of pesticide applied to crops, and thereby reduce potentially deleterious effects on the environment and human health. Although forecasts may have additional objectives, the three listed here are commonly cited as justification for developing and applying forecasts.

Forecasts are useful in some situations, but not others. They are appropriate only for those diseases whose important occurrence is sporadic (Bourke, 1970). If the disease is always important, management efforts will always be required, and the forecast provides no useful information. If a disease is never important, effort spent in forecasting is unnecessary. Effective technology for disease suppression and mechanisms to distribute the results of forecasts are necessary to justify the development or use of a forecast (Bourke, 1970). Unfortunately, the appropriateness of a forecast is not always accurately known prior to its application. It was only after several seasons of experimenting with a weather-sensitive potato late blight forecast that researchers realised that the macroclimate of the potato growing regions of Long Island, New York, USA, was consistently favourable to late blight. The forecast provided little useful information.

Existing forecasts use diverse elements because the diseases of interest are biologically and epidemiologically diverse and the technology of disease suppression is diverse. Study of the biology of the pathogen serves to identify the phase of the life cycle most readily influenced by available technology. As technology changes, a different phase may become the one most readily influenced by the new technology. The epidemiology of the host–pathogen interaction provides general guidelines for a disease forecast. For diseases in which variation in the amount of initial inoculum contributes significantly to variation in disease occurrence, a forecast should include some assessment of initial inoculum. For diseases in which variation in efficacy of initial inoculum contributes significantly to variation in disease occurrence, assessments of the efficacy of initial inoculum should be included in a forecast. Finally, for diseases in which variation in rapidity of pathogen secondary cycles contributes significantly to variation in disease occurrence, consideration of the rapidity of secondary cycles needs to be included in the forecast.

Some of the forecasts developed during the last 50 years have consistently provided useful information to growers and disease managers. Although these forecasts were developed without formal mathematical modelling, they are consistent with the epidemiological principles of forecasting.

Several successful forecasts rely on estimates of the amount of initial inoculum (Table I). The forecast for Stewart's wilt of corn is particularly interesting because it was developed before the epidemiology of the pathogen was fully known. From observations of a positive correlation between mild winters and severe disease, Stevens (1934) predicted that disease would be severe if the sum of the monthly mean temperatures for December, January and February was $>100°F$. Disease would be less severe if this sum was lower and disease would be mild if the sum was $< 90°F$. At about the same time, other workers discovered that the bacteria overwintered in the corn flea beetle (Elliott and Poos, 1934), which explained the efficacy of the forecast. Fewer beetles survived in cold winters than in mild ones. Thus, the forecast predicted the relative size of the overwintering beetle population, and therefore the bacteria population. The forecast is still used with only minor modification (Castor *et al.*, 1975).

Other forecasts rely on estimates of the efficacy of the initial inoculum (Table I). One of the best known of these is the forecast for apple scab, caused by *Venturia inaequalis*. In many apple-growing regions, *V. inaequalis* ascospores (initial inoculum) are consistently abundant in the spring, but scab is not always severe because the efficacy of the ascospores depends on environmental conditions. Because of the potential damage, large amounts of fungicide are used to suppress scab. In the northeastern United States, half of the fungicide applied by farmers is used to prevent scab (Andrilenas, 1974). Only juvenile and recently matured tissues are susceptible (Boone, 1971). The pathogen requires 1–3 days after inoculation to colonise tissue so that it is sensitive to fungicides which partially penetrate the cuticle. Disease severity is associated with periods of leaf wetness while leaves are actively growing in the spring. Mills (1944) identified the relationships between temperature and duration of leaf wetness that resulted in severe, moderate or light infection. The severity of disease increased with increasing length of leaf wetness periods and with increasing temperature (up to 22°C). In many apple-growing regions, the occurrences of light, moderate and heavy infection periods are widely and rapidly publicised after their occurrence. If trees were unprotected by fungicide during the infection period, growers can apply a fungicide capable of partially penetrating the cuticle.

For many polycyclic diseases, forecasts have been based on the rapidity of secondary cycles or on the amount of secondary inoculum (Table I). Forecasts are made during the growing season, and usually indicate the timing of the initial spray application of the season and/or the frequency of sprays during the season. Various criteria have been used as the basis for decisions in these forecasts. In some cases weather factors are used; in other

Table I. Some plant disease forecasting schemes.

Crop	Disease	Epidemiological Emphasis[a]	Pathogen	References
Apple	Scab	R	*Venturia inaequalis*	Mills (1944), Jones *et al.* (1980)
Barley	Powdery mildrew	r	*Erysiphe graminis hordei*	Channon (1981)
Celery	Early blight	X	*Cercospora apii*	Berger (1969a,b)
Cherry	Leafspot	R, r	*Coccomyces heimalis*	Eisensmith and Jones (1981)
Grape	Downy mildew	r	*Plasmopara viticola*	Tarr (1972)
Lima bean	Downy mildew	r	*Phytophthora phaseoli*	Hyre (1957)
Maize	Stewart's wilt	X_0	*Erwinia stewartii*	Stevens (1934)
Maize	Leaf blight	r	*Helminthosporium turcicum*	Berger (1970)
Onion	Leaf blight	X	*Botrytis squamosa*	Shoemaker and Lorbeer (1977)
Pea	Root rot	X	*Aphanomyces euteiches*	Reiling *et al.* (1960), Sherwood and Hagedorn (1958)
Pear	Fireblight	X	*Erwinia amylovora*	Thomson *et al.* (1977, 1982)
Peanut	Leafspot	r	*Cercospora* sp.	Jensen and Boyle (1965, 1966), Parvin *et al.* (1974)
Potato	Early blight	X	*Alternaria solani*	Harrison *et al.* (1965)
Potato	Late blight	r	*Phytophthora infestans*	Beaumont (1947), Fry *et al.* (1983), Grainger (1953), Hyre (1954), Krause *et al.* (1975), Schrödter and Ullrich (1966), Wallin (1962)
Rice	Wet stem rot	X_0	*Sclerotium oryzae*	Webster (1974)
Sugar beet	*Sclerotium* rot	X_0	*Sclerotium rolfsii*	Backman *et al.* (1981), Leach and Davey (1938)
Sugar beet	Yellows	X, X_0	Beet yellows virus; beet mild yellows virus	Hull (1968, 1976), Watson *et al.* (1975)
Tea	Blister blight	X, r	*Exobasidium vexans*	DeWeille (1960)
Tobacco	Blue mold	r	*Peronospora tabacina*	Miller (1958)
Tomato	Early blight	r	*Alternaria solani*	Madden *et al.* (1978)
Wheat	Stem rust	r, X	*Puccinia graminis tritici*	Eversmeyer *et al.* (1973)
Wheat	Stripe rust	r, X	*Puccinia striiformis*	Rabbinge and F. H. Rijsdijk (personal communication)
Wheat	Leaf rust	r, X	*Puccinia recondita*	Burleigh *et al.* (1972)

[a] Epidemiological emphasis may be r (rate of secondary cycles), R (efficacy of initial inoculum), X_0 (amount of initial inoculum or initial disease) or X (disease or inoculum threshold).

cases assessments of inoculum are used; and in some cases both weather and inoculum assessments are included. Forecasts of potato late blight, induced by *Phytophthora infestans*, and of *Cercospora* leaf spot of peanuts, induced by *Cercospora* spp., depend on weather factors (Barriga *et al.*, 1961; Beaumont, 1947; Grainger, 1953; Hyre, 1954; Jensen and Boyle, 1966; Krause *et al.*, 1975; Nugent, 1950; Schrödter and Ullrich, 1966; Van Everdingen, 1926; and Wallin, 1962). Forecasts of celery early blight (induced by *Cercospora apii*), of wheat leaf and stem rusts (induced by *Puccinia recondita* and *Puccinia graminis tritici*, respectively, and of potato early blight (induced by *Alternaria solani*) depend on assessments of air-borne pathogen conidia (Berger, 1969a,b; Burleigh *et al.*, 1972; Harrison *et al.*, 1965). Forecasts for sugar beet yellows (induced by beet yellows and beet mild yellowing viruses interacting with aphid vectors) depend on weather factors and assessments of vector populations (Hull, 1968, 1976; Hurst, 1965; and Watson *et al.*, 1975).

III. CONTRIBUTIONS OF MODELLING TO FORECASTING

The extensive use of mathematical models in plant pathology is just beginning. The initial excitement of computer disease simulators has matured as scientists investigate the role of such models in the science and practice of plant pathology. Pathologists are now beginning to employ such models as tools to help investigations of biology as well as tools for making management decisions. In this section we investigate how models have been used to streamline the use of forecasts, to develop forecasts and to assess the value of forecasts.

A. Mathematical Models and Computers for Implementing Forecasts

In the United States computers have streamlined the use of several different forecasts. Probably the best known of these is BLITECAST, a computerised forecast for potato late blight developed by Krause and colleagues at Pennsylvania State University (Krause *et al.*, 1975; Krause and Massie, 1975). BLITECAST combined a potato late blight forecasting system developed by Hyre (1954) with a forecast system developed by Wallin (1962). Hyre's system forecasts late blight 7–14 days after 10 consecutive rain-favourable days. Wallin's system is based on the occurrence of periods of high relative humidity. 'Severity values' are assigned to periods of high relative humidity. Severity values increase to a maximum of 4/day with increases in the duration of high relative humidity and with increases of temperature to

26.6°C. The initial appearance of late blight is predicted after the accumulation of 18–21 severity values from the time of plant emergence, and further occurrences are predicted after each subsequent accumulation of three severity values. BLITECAST was initially programmed on a large main-frame computer, but subsequently has become available on a microcomputer coupled with a weather data logger (MacKenzie, 1981). The microcomputer-data logger is a self-contained unit and can be placed into a potato field.

Other well-established forecasts have been programmed onto computers. The method of determining the potential for Stewart's wilt has been programmed onto a main-frame computer (Castor *et al.*, 1975). The method for determining apple scab infection periods has also been programmed onto computers. This program has been incorporated with a weather data logger into a self-contained unit (Jones *et al.*, 1980). The definition of infection periods was changed slightly as the apple scab forecaster was developed.

B. Mathematical Modelling for Developing Disease Forecasts

Several disease forecasts have evolved from the analysis of disease using computer simulators. Forecasts for tomato early blight induced by *A. solani*, for potato late blight, and for stripe rust of wheat, induced by *Puccinia striiformis*, will be considered here.

A computerised Forecasting system for *A. solani* on Tomato (FAST) was developed with the aid of a computer simulator of *A. solani* on tomatoes (Waggoner and Horsfall, 1969). The goal of FAST is efficient scheduling of fungicide applications (Madden *et al.*, 1978). Two different models (*S* and *R*), used to determine the need for fungicide applications, were synthesized from known relationships between *A. solani* and the microenvironment. Model *S* is based on the duration and temperature of periods of leaf wetness. Model *R* is based on temperature, relative humidity and rainfall during the previous 5–7 days. A fungicide application is needed only when transplanting occurred at least 28 days earlier and *S* values total at least 35. If the 7-day *S* value is greater than or equal to 13, or if the 5 day *R* value is greater than 8, the grower should apply fungicide every 7-days. If the 7-day *S* value is greater than 12 and the 5-day *R* value is greater than 8, the grower should apply fungicide every 5 days. FAST requires the following environmental monitoring: temperature, relative humidity, leaf wetness and rainfall. Use of a microcomputer programmed with FAST allows easy determination of the appropriate schedule for fungicide applications. Initial tests indicate that FAST enables efficient use of fungicide for suppression of *A. solani* (Pennypacker *et al.*, 1983).

A procedure for timing fungicide applications to suppress potato late blight was developed using two simulation models. The first model simulated the effects of environment, fungicide and host resistance on *P. infestans* in potato foliage (Bruhn and Fry, 1981). Using weather data for three successive growing seasons, the model accurately predicted epidemic development in cultivars with different levels of rate-reducing resistance. The second model described fungicide deposition and subsequent redistribution on potato foliage (Bruhn and Fry, 1982a,b). The fungicide model was based on analysis of chlorothalonil deposition and redistribution, but subsequently has been generalised to include effects of triphenyltin hydroxide and captafol (Spadafora *et al.*, 1984). When the fungicide model was linked to the pathogen model, the combination accurately reflected the effects of fungicides, weather and host resistance on epidemic development (J. A. Bruhn and W. E. Fry, unpublished results).

The first step in using the pathogen and fungicide models to time fungicide applications was to develop decision rules for timing applications according to weather. The degree to which weather was favourable for late blight development was quantified in terms of blight units. These units were determined from analysis using the pathogen simulator. A series of simulations were performed at various combinations of temperature and daily periods of high relative humidity (>90%). The results for a cultivar of a specific level of resistance were quantified in terms of area under the disease progress curve (AUDPC) (Fig. 1). Blight units were obtained by dividing values for AUDPC by five (Table II). Blight units accumulate more slowly for moderately resistant cultivars than for susceptible cultivars (Table II).

The second step in developing a procedure for timing fungicide applications was to develop decision rules based on fungicide residue. Depletion of

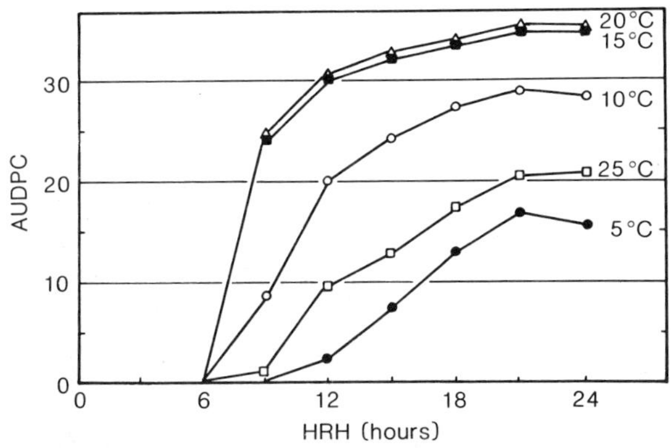

Fig. 1. Influence of temperature and periods of high relative humidity (≥90%) on potato late blight development on a susceptible cultivar as determined by analysis of computer simulation models. Courtesy of J. A. Bruhn and W. E. Fry (unpublished).

Table II. Blight units as determined by cultivar resistance, temperature and periods of high relative humidity ($\geq$90%)[a].

Temperature (°C)	Cultivar Resistance	Hours of Relative Humidity $\geq$ 90%							
>27	Suscept.	24							
	Mod. suscept.	24							
	Mod. resist.	24							
23–27	Suscept.	6	7–9	10–12	13–15	16–18	19–24		
	Mod. suscept.	9	10–18	19–24					
	Mod. resist.	15	16–24						
13–22	Suscept.	6					7–9	10–12	13–24
	Mod. suscept.	6	7	8	9	10	11–12	13–24	
	Mod. resist.	6	7	8	9	10–12	13–24		
8–12	Suscept.	6	7	8–9	10	11–12	13–15	16–24	
	Mod. suscept.	6	7–9	10–12	13–15	16–18	19–24		
	Mod. resist	9	10–12	13–15	16–24				
3–7	Suscept.	9	10–12	13–15	16–18	19–24			
	Mod. suscept.	12	13–24						
	Mod. resist.	18	19–24						
<3	Suscept.	24							
	Mod. suscept.	24							
	mod. resist.	24							
Blight units[b]:		0	1	2	3	4	5	6	7

[a] From Fry *et al.* (1983). Key: suscept., susceptible; mod., moderately; resist., resistant.
[b] Blight units are given for each column.

fungicide from the potato canopy was quantified in terms of fungicide units from analysis of the fungicide simulation models. For chlorothalonil the most important factors influencing fungicide depletion were rainfall and time since application (Table III).

The need for an additional fungicide spray is indicated by the accumulation of a minimum number of blight units or a minimum number of fungicide units (Table IV). The minima were determined through simulation analysis. Simulations were done using five seasons of diverse weather for a range of thresholds for blight units and a range of thresholds for fungicide units. The values of blight units and fungicide units which minimised costs (crop loss from disease and cost of fungicide applications) were the values adopted in the decision rules (Table IV). Host resistance was incorporated into these decision rules by allowing greater accumulation of blight-favourable weather and further depletion of fungicide before fungicide is applied on moderately resistant cultivars than is allowed for susceptible ones (Table IV).

In field experiments the simulation forecast enabled efficient disease management. Late blight was suppressed similarly on susceptible and resistant cultivars, but resistant cultivars received less fungicide than did susceptible ones (Fry *et al.*, 1983).

The two simulation models were also used to identify the most appropriate fixed interval for applications in upstate New York for cultivars with different levels of resistance. Simulations were done using data for weather typical of upstate New York, and for susceptible, moderately susceptible or moderately resistant cultivars. For susceptible, moderately susceptible and moderately

Table III. Fungicide units for chlorothalonil as determined by rainfall and time since last fungicide application.[a]

Days since Fungicide Application	Daily Rainfall (mm)						
1	<1	—	—	1	2–3	4–6	> 6
2	<1	—	1	2–4	5–8	>8	—
3	<1	—	1–2	3–5	>5	—	—
4–5	<1	—	1–3	3–8	>8	—	—
6–9	<1	—	1–4	>4	—	—	—
10–14	<1	1	2–8	>8	—	—	—
>14	<1	1–8	>8	—	—	—	—
Fungicide units[b]:	1	2	3	4	5	6	7

[a] From Fry *et al.* (1983).

[b] Fungicide units are given for each column.

Table IV. Decision rules for the simulation forecast for potato late blight.[a]

	Cultivar Resistance	
	Susceptible	Moderately Resistant
Fungicide should be applied if fungicide has not been applied within 5 days		
AND cumulative blight units since last spray exceed:	30	40
OR cumulative fungicide units since last spray exceed:	15	25

[a] From Fry *et al.* (1983).

resistant cultivars, the application intervals which minimised average costs and which limited variance of costs were 7, 9 and 10–12 days, respectively. Thus, in the event that a grower elects not to monitor weather, the frequency of fungicide applications can still be adjusted to complement cultivar resistance.

A computer simulation model of *P. striiformis* on wheat has helped in the development of decision rules useful in determining the need for applying fungicide to suppress stripe rust on wheat in the Netherlands. This decision rule is part of a larger pest management system: EPIPRE (EPIdemics, PREdiction and PREvention) (R. Rabbinge, personal communication). Detailed models of stripe rust were used to develop simple growth models. Disease development is assumed to be exponential: $X = X_0[\exp(rt)]$. The growth models of EPIPRE are solved for specific fields (R. Rabbinge and F. H. Rijsdijk, personal communication). The necessary information provided by the grower includes disease intensity, growth stage of crop, culture, nitrogen fertilisation, region of the country, sowing time, some soil characteristics and herbicide and growth regulator application. This information plus detailed weather data are used to predict disease development (J. C. Zadoks, F. H. Rijsdijk and R. Rabbinge, personal communication). Predicted disease development is then compared to damage thresholds. The economics of treating a specific field are then determined by computers and a recommendation to apply or not to apply fungicide is made.

EPIPRE is apparently quite successful. In addition to *P. striiformis*, the system includes advice on treatments for brown rust (*P. recondita*), wheat powdery mildew (*Erysiphe graminis*), *Septoria* spp., and the aphids *Sitobion avenae*, *Metopolophium dirhodum* and *Rhopalosiphum padi*. By 1981 in the

Netherlands there were 1155 fields included in the system (R. Rabbinge and F. H. Rijsdijk, personal communication).

C. Mathematical Models for Assessing the Value of Forecasts

Mathematical models are potentially valuable tools for evaluating the benefits of forecast systems. We describe here simulation experiments done as an initial evaluation of the economics of using BLITECAST to forecast potato late blight. For this evaluation, the effectiveness of applications timed according to BLITECAST was compared with applications at regular time intervals (Fohner *et al.*, 1982, and unpublished).

The effect on farm income from using BLITECAST instead of applications at regular intervals may depend on the amount of inoculum to which the potatoes are exposed and the favourability of the environment for late blight. Consequently, a thorough evaluation of the performance of BLITECAST requires it to be compared with spraying at regular intervals for a variety of environments and levels of inoculum. Locating or creating this variety of conditions in farmers' fields or experimental plots would be costly, so computer simulation may be a useful and cost-effective supplement to field tests for comparing BLITECAST and regular spraying.

The comparisons were made using the two simulation models described previously. The first simulated pathogen development according to weather, host resistance and fungicide amount (Bruhn, 1981; Bruhn and Fry, 1981). The second described fungicide deposition and redistribution as a function of weather (Bruhn, 1981; Bruhn and Fry, 1982a,b). The comparison was made for the potato cultivar 'Katahdin'.

The simulation model represented the development of late blight of 50 m^2 of potato foliage. In addition to modelling the production and dissemination of inoculum within this plot, the model included the influx of inoculum into the plot from exogenous sources. This influx, which initiated and sustained the epidemic in the plot, began on day 50 of the 120-day growing season. The daily influx of inoculum varied according to the favourability of the weather for late blight. The favourability of the weather was rated in 'blight units' (Table II). BLITECAST and regular intervals were compared at three levels of inoculum. At the low level the influx per day was 2^{BU}, (i.e. 2 raised to the power equal to the blight units for that day). At the moderate exposure, influx was 6^{BU}; at the high exposure, influx was 10^{BU}.

BLITECAST and regular sprayings were compared for 10 seasons using macroclimate rainfall and temperature data recorded at Geneva, New York. Periods of high relative humidity ($\geq 90\%$) in the crop canopy (microclimate) were estimated using a stochastic function derived from such measurements in field plots of potatoes during three seasons. The function was derived by regressing hours of high relative humidity on minimum daily temperature, occurrence of rain on the current day and occurrence of rain on the previous day. Data for hours of high relative humidity ($\geq 90\%$) in the crop canopy were calculated for three microclimates. Data for a microclimate favourable for late blight were calculated with a stochastic function estimated from measurements of rainfall, temperature and relative humidity at a location favourable for the disease. Data for a moderately favourable microclimate were calculated by subtracting 1 hour of high relative humidity from the values for the favourable microclimate. Data for an unfavourable microclimate were calculated by subtracting 2 hours.

BLITECAST has two parts, one for determining when fungicide should be first applied in a season, and the other for scheduling subsequent applications. Both parts are based on an assumed relationship between weather and the need to apply fungicide to control late blight. This simulation experiment was not appropriate for testing the first part of BLITECAST because the influx of inoculum began on day 50 regardless of the preceding weather. Only the effectiveness of BLITECAST at scheduling subsequent applications was tested. For both BLITECAST and the applications at regular intervals, the first fungicide application was made on day 50. No fungicide was applied after day 116 because later applications would have had no effect on observed disease because of the length of the latent period.

To schedule applications of fungicide using BLITECAST, data for relative humidity, temperature and rainfall were evaluated once every 4 days, beginning on day 55, to select one of four options: no applications of fungicide during the time before the next evaluation of microclimate data, reevaluation of microclimate data in 2 days, application of fungicide on a 7-day interval, and application on a 5-day interval. When reevaluation of the data in 2 days was recommended, the additional evaluation was made without affecting the regular 4-day schedule of evaluations.

1. Comparing the Performance of BLITECAST with the Fungicide Schedule Currently Used by Growers

The average number of fungicide applications per season and the average percent defoliation from disease at the end of the season are summarised in Table V for BLITECAST and a 7-day interval of application, the interval of application commonly recommended for and used by potato growers in the state of New York.

For the microclimate favourable for late blight, the use of BLITECAST resulted in a higher average number of fungicide applications than did application at 7-day intervals. With low exposure to inoculum, both schemes resulted in very low average defoliation. With high or moderate exposure to

Table V. Average defoliation from late blight and average number of fungicide applications for 10 simulated seasons for spraying according to BLITECAST and at 7-day intervals.[a]

	Microclimate					
	Favourable		Moderately Favourable		Unfavourable	
Average Number of Fungicide Applications per Season:	BLITECAST	7-day	BLITECAST	7-day	BLITECAST	7-day
	10.6	10	8.7	10	6.6	10
Exposure to Inoculum	Average Percent Defoliation					
High	40.1 (7.9)[b]	35.2 (8.2)	34.1 (8.6)	21.2 (8.5)	23.2 (8.9)	10.9 (5.2)
Moderate	16.3 (6.0)	15.2 (6.7)	16.5 (9.1)	10.0 (6.1)	9.5 (4.6)	3.9 (3.4)
Low	0.2 (0.1)	0.4 (0.3)	1.0 (0.7)	0.2 (0.2)	0.1 (0.0)	0.0 (0.0)

[a] Data are from Fohner *et al.* (1984).
[b] Values in parentheses are standard deviation of average defoliation.

inoculum, routine applications at 7-day intervals resulted in less average defoliation than did applications made according to BLITECAST.

For the moderately favourable microclimate, the use of BLITECAST resulted in an average of 1.3 fewer applications of fungicide than the 7-day interval. For the unfavourable microclimate, the BLITECAST scheme resulted in an average of 3.4 fewer fungicide applications than the weekly applications. When exposure to incoming inoculum was low, ranging daily according to weather from 1 to 64 spores, these reductions in fungicide use for the moderately favourable and unfavourable microclimates were realised with virtually no increase in average disease. For moderate and high exposure to inoculum, however, the reductions were associated with an increase in average defoliation.

The costs associated with different levels of defoliation must be known to determine the effects of the different fungicide timing techniques on net farm income. The 7-day interval, with a lower disease level but a higher cost of fungicide, may have an effect similar to or different from BLITECAST on net farm income. The most common costs from late blight in upstate New York are probably from infection of tubers and increases in production costs in response to detection of the disease. Defoliation from the disease rarely becomes extensive enough in New York to affect tuber growth significantly (Fohner *et al.*, 1982). Infection of tubers directly reduces marketable yield at harvest and can result in large additional losses during storage. Production costs may be increased by late blight because, upon detecting the disease, growers generally increase the dosage and frequency of fungicide applications, and may hire aerial application to improve timeliness. Changing from prophylactic fungicides to more expensive therapeutic fungicides is another new option for responding to detection of late blight. In addition to these changes in management of the crop in the field, detection of late blight may lead to more intensive inspection of the crop during harvest and grading, and to changes in storage and marketing. The relationship between amount of defoliation and the number and cost of infected tubers, and the relationship between amount of defoliation and changes in management practices are both unknown.

The task of relating defoliation in an experiment to cost in a commercial potato field is further complicated by the differences between experimental and commercial conditions. One obvious difference between the two conditions is the change in scheduling of fungicide applications that often occurs when late blight is detected in commercial potato fields, compared to the lack of change in the experiment regardless of the severity of disease. Another difference between experimental and commercial conditions is the amount of inoculum to which the potato foliage is exposed. In the simulation experiment, exposure to the moderate and high inoculum resulted in detectable defoliation far more frequently than is observed for New York potato farms. It is likely that the high and moderate exposures in the simulation experiment exceed those common in commercial potato fields. The exposure to inoculum in commercial potato fields, however, is unknown.

2. Evaluating BLITECAST and Regular Intervals when Costs of Disease Cannot Be Estimated

Because the relationships are unknown between defoliation and cost, and between experimental and commercial conditions, the costs of late blight when applying fungicide according to BLITECAST or at 7-day intervals cannot be estimated directly from defoliation in the simulation experiment. Without those estimates of cost, the combined cost of late blight and fungicide applications for BLITECAST and the regular application cannot be compared. It is possible, however, to use simulation to compare BLITECAST with a range of regular intervals of fungicide application, and thereby test whether any of the regular intervals can control disease as effectively as with BLITECAST with fewer applications. The average defoliation and number of fungicide applications for BLITECAST and range of regular

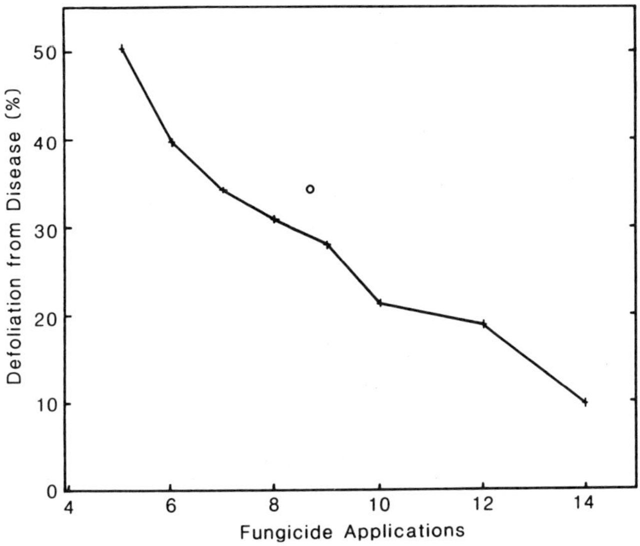

Fig. 2. Average defoliation from late blight and average number of fungicide applications for 10 simulated seasons for BLITECAST and a range of regular intervals. Microclimate, moderate; exposure to inoculum, high; ○, BLITECAST; +, regular intervals.

intervals, resulting in from 5 to 14 applications per season, are plotted in Fig. 2 for the moderately favourable microclimate and high exposure to inoculum. Applying fungicide at an interval of once every 10 days (7 applications) resulted in the same average defoliation from late blight as BLITECAST, while resulting in an average of 1.7 fewer applications of fungicide per season. Results for the other microclimates and exposures to inoculum also indicated that regular intervals could control late blight as effectively as BLITECAST and with fewer applications of fungicide, although the magnitude of defoliation differed substantially among the microclimates and exposures to inoculum.

The comparison between BLITECAST and the range of regular intervals is useful evidence that BLITECAST does not improve the timing of fungicide applications compared to regular intervals, but it has limitations for helping potato growers choose between the use of BLITECAST and a particular regular interval of fungicide application. The trade-off between disease and fungicide applications varies among microclimates and exposure to inoculum, so the optimal regular interval and the optimal choice between BLITECAST and a particular interval may vary among locations. BLITECAST may be less effective than regular intervals in general, but may be better than a particular interval for a particular location.

3. Conclusions from the Simulation Experiment

Although the optimal trade-off between disease and fungicide applications could not be identified from the simulation experiment, conclusions can be drawn about the relative performance of BLITECAST and application of fungicide at regular intervals. For microclimates favourable for late blight, the strategy of routine application at 7-day intervals was better than BLITECAST. For moderately favourable microclimates and low exposure to inoculum, the use of BLITECAST reduced fungicide applications while controlling disease as well as did the routine application at 7-day intervals, but the small savings in fungicide probably would not justify the cost and inconvenience of using BLITECAST. When exposure to inoculum was moderate or high in the moderately favourable microclimate, the potential value of using BLITECAST was further diminished because disease was higher than with the routine, 7-day application. For unfavourable microclimates, the value of applying fungicide according to BLITECAST could not be compared with the regular weekly application because the relationship between defoliation in the experiment and costs in commercial conditions was unknown. Comparison of the use of BLITECAST with a range of regular intervals of applications, however, indicated that increasing the length of regular intervals would be better than using BLITECAST for reducing the use of fungicide in microclimates that are not favourable for late blight. These conclusions, of course, depend on the realism of the models of late blight, fungicide and microclimate used in the experiment.

IV. CONCLUSIONS

During the last 10 years mathematical modelling in plant pathology has matured from an initial phase of only constructing models to this current phase of applying them. Models have been used successfully in on-line pest management activities (FAST, Stewart's wilt, EPIPRE) and they have been used to help develop predictive systems (potato late blight forecast, EPIPRE). They continue to be useful in theoretical studies of the biology of plant disease. Their initial application to the economics of pest management has provided startling predictions which warrant careful consideration. These contributions generate much enthusiasm for models. Their contribution to the practice and science of plant pathology is certain to increase.

ACKNOWLEDGEMENTS

This chapter has benefited from the efforts of several individuals. We wish to thank D. Lane and M. Brodie for typing drafts of the manuscript and A. E. Apple for reviewing calculations and narrative for preparing figures.

REFERENCES

Andrilenas, P. A. (1974). *Agric. Econ. Rep.* **252**, Econ. Res. Serv., U.S. Dept. Agric., Washington, D.C.

Backman, P. A., Rodriguez-Kabana, R., Caulin, M. C., Beltramini, E. and Ziliani, N. (1981). *Plant Dis.* **65**, 419–421.

Barriga, R. O., Thurston, H. D. and Heidrick, L. E. (1961). *Agric. Trop.* **17**, 617–622.

Beaumont, A. (1947). *Trans. Br. Mycol. Soc.* **31**, 45–53.

Berger, R. D. (1969a). *Proc. Fla. State Hortic. Soc.* **82**, 107–111.

Berger, R. D. (1969b). *Phytopathology* **59**, 1018 (Abstr.).

Berger, R. D. (1970). *Phytopathology* **60**, 1284 (Abstr.).

Boone, D. M. (1971). *Annu. Rev. Phytopathol.* **9**, 297–318.

Bourke, P. M. A. (1970). *Annu. Rev. Phytopathol.* **8**, 345–370.

Bruhn, J. A. (1981). Ph.D. Thesis, Cornell Univ., Ithaca, New York.

Bruhn, J. A. and Fry, W. E. (1981). *Phytopathology* **71**, 612–616.

Bruhn, J. A. and Fry, W. E. (1982a). *Phytopathology* **72**, 1301–1305.

Bruhn, J. A. and Fry, W. E. (1982b). *Phytopathology* **72**, 1306–1312.

Burleigh, J. R., Eversmeyer, M. G. and Roelfs, A. P. (1972). *Phytopathology* **62**, 947–953.

Castor, L. L., Ayers, J. E., MacNab, A. A. and Krause, R. A. (1975). *Plant Dis. Rep.* **59**, 533–536.

Channon, A. G. (1981). *Ann. Appl. Biol.* **97**, 43–53.

DeWeille, G. A. (1960). *Neth. J. Agric. Sci.* **8**, 183–210.

Eisensmith, S. P. and Jones, A. L. (1981). *Plant Dis.* **65**, 955–958.

Elliott, C. and Poos, F. W. (1934). *Science* **80**, 289–290.

Eversmeyer, M. G., Burleigh, J. R. and Roelfs, A. P. (1973). *Phytopathology* **63**, 348–351.

Fohner, G. R., White, G. B. and Fry, W. E. (1982). Econ. Rep. 82–38. Dept. Agric. Econ., Cornell Univ., Ithaca, New York.

Fohner, G. R., White, G. B. and Fry, W. E. (1984). *Phytopathology* **74**, 1145–1147.

Fry, W. E., Apple, A. E. and Bruhn, J. A. (1983). *Phytopathology* **73**, 1054–1059.

Grainger, J. (1953). *Nature* (*London*) **171**, 1012–1014.

Harrison, M. D., Livingston, C. H. and Oshima, N. (1965). *Am. Potato J.* **42**, 319–327.

Hull, R. (1968). *Plant Pathol.* **17**, 1–10.

Hull, R. (1976). *Ann. Appl. Biol.* **82**, 1–10.

Hurst, G. W. (1965). *Plant Pathol.* **14**, 47–53.

Hyre, R. A. (1954). *Plant Dis. Rep.* **38**, 245–253.

Hyre, R. A. (1957). *Plant Dis. Rep.* **41**, 7–9.

Jensen, R. E. and Boyle, L. W. (1965). *Plant Dis. Rep.* **49**, 975–978.

Jensen, R. E. and Boyle, L. W. (1966). *Plant Dis. Rep.* **50**, 810–814.

Jones, A. L., Lillevik, S. L., Fisher, P. D. and Stebbins, T. C. (1980). *Plant Dis.* **64**, 69–72.

Krause, R. A. and Massie, L. B. (1975). *Annu. Rev. Phytopathol* **13**, 31–47.

Krause, R. A., Massie, L. B. and Hyre, R. A. (1975). *Plant Dis. Rep.* **59**, 95–98.

Leach, L. D. and Davey, A. E. (1938). *J. Agric. Res.* **56**, 619–631.

MacKenzie, D. R. (1981). *Plant Dis.* **65**, 394–399.

Madden, L., Pennypacker, S. P. and MacNab, A. A. (1978). *Phytopathology* **68**, 1354–1358.

Miller, P. R. (1958). *In* "Plant Pathology: Problems and Progress 1908–1958," pp. 557–565. Univ. of Wisconsin Press, Madison.

Mills, W. D. (1944). *Cornell Extension Bull.* **630**.

Nugent, T. J. (1950). *Plant Dis. Rep. Suppl.* **190**, 9–13.

Parvin, D. W. Jr., Smith, D. H. and Crosby, F. L. (1974). *Phytopathology* **64**, 385–388.

Pennypacker, S. P., Madden, L. V. and MacNab, A. A. (1983). *Plant Dis.* **67**, 287–289.

Reiling, T. P., King, T. H. and Fields, R. W. (1960). *Phytopathology* **50**, 287–290.

Schrödter, H. and Ullrich, J. (1966). *Phytopathol. Z.* **56**, 265–278.
Sherwood, R. T. and Hagedorn, D. J. (1958). *Wis., Agric. Exp. Stn. Bull.* **531**.
Shoemaker, P. B. and Lorbeer, J. W. (1977). *Phytopathology* **67**, 409 (Abstr.).
Spadafora, V. J., Bruhn, J. A. and Fry, W. E. (1984). *Phytopathology* **74**, 519–523.
Stevens, N. E. (1934). *Plant Dis. Rep.* **12**, 141–149.
Tarr, S. A. J. (1972). "Principles of Plant Pathology." Winchester, New York.
Thomson, S. V., Schroth, M. N., Moller, W. J., Reil, W. O., Bentel, J. A. and Davis, C. S. (1977). *Calif. Agric.* **31**, 12–14.
Thomson, S. V., Schroth, M. N., Moller, W. J. and Reil, W. O. (1982). *Plant Dis.* **66**, 576–579.
Van Everdingen, E. (1926). *Tijdschr. Plantenziekten* **32**, 129 (Abstr.).
Waggoner, P. E. and Horsfall, J. G. (1969). *Bull. Conn. Agric. Exp. Stn. New Haven* **698**.
Wallin, J. R. (1962). *Am. Potato J.* **39**, 306–312.
Watson, M. A., Heathcote, G. D., Lauckner, F. B. and Sowray, P. A. (1975). *Ann. Appl. Biol.* **81**, 181–198.
Webster, R. K. (1974). *Proc. Am. Phytopathol. Soc.* **1**, 106–107.

8

CONSTRUCTION OF PREDICTIVE MODELS: II. FORECASTING CROP LOSSES

P. S. Teng

Department of Plant Pathology, University of Minnesota, St. Paul, Minnesota, USA

I. INTRODUCTION

Crop loss assessment has received increased attention as a legitimate area of research in plant protection largely because of the use of mathematical techniques, foremost of which is modelling. Zadoks (1972) suggested that 'every respectable science goes quantitative'; with loss assessment, modelling has given respectability to a previously subjective area of study and has speeded its development into a science. In reality, modern loss assessment depends on the development of models that can depict the interaction between pathogen, host and environment in causing loss. Extensive reviews of the subject have previously been made, and these generally cover topics additional to crop loss models (Chester, 1950; Large, 1966; James, 1974; James and Teng, 1979). In this chapter, the focus will be on predictive models for loss forecasting, in particular, the biological and statistical bases for such models.

A. Terminology

The first term that is of concern in this chapter is 'model', which is generally understood by modellers and non-modellers alike, but which is difficult to define precisely to satisfy the many uses of the term. A model is a representation of a defined system; mathematical models represent a system by symbols (Teng, 1981b). In this chapter, a model is considered as any symbolic representation of a disease–loss system capable of loss prediction, given some input parameter such as disease severity. This broad interpretation of a model is used to include simple equations as well as complex simulation models.

All models have a predictive ability to which some confidence level can be assigned. Prediction implies a quantitative output when a model is given an input value(s), which may represent an event that has already occurred or is expected to occur in the future. Forecasting suggests a projection into the future. As Jones (1983) has noted, there are few forecasting systems for diseases; the majority are predictive, making use of what is known of the system to estimate current system behaviour, or at best, a short time into the future.

A distinction is explicitly made in this chapter between yield loss and crop loss, following the convention of MacKenzie (1983) and Teng and Oshima (1983). Yield loss is loss attributable to one causal factor while crop loss refers to the multiple factor system. There is an interplay between size of the spatial unit for which losses are to be assessed and yield/crop loss. Plot-sized units commonly enable precise definition of yield loss, while for a county-sized unit, crop loss would be the entity assessed. Crop loss has been generally considered the reduction in yield relative to a reference yield, e.g. crop loss = attainable yield − actual yield (Chiarappa, 1971), with the yield levels clearly defined (Zadoks and Schein, 1979). Loss may also be considered any measurable reduction in quality or quantity of crop produce (Chiarappa, 1971).

Disease may be quantified as either disease incidence or disease severity, in proportion terms (James and Teng, 1979). Disease intensity is used to mean either incidence or severity, as suggested in the Food and Agricultural Organisation (FAO) crop loss manual (Chiarappa, 1971). Incidence (I) is the number of plant units infected, as a proportion of the total number of units assessed, while severity (S) is area of plant tissue affected by disease, expressed as a proportion of the total area.

B. Objectives of Loss Forecasting

It has been argued that the science of plant pathology started because of devastating crop losses (James, 1974; Main, 1983). It has been further argued that, in spite of this, only a few reliable estimates of field losses caused by plant diseases are available today (James, 1980). What then is the rationale for crop loss assessment? As agriculture becomes increasingly resource constrained, whether in actual production or in research, it has become apparent that we need to justify and rationalise the use of inputs in crop management. Crop loss information provides a means for evaluating the potential benefits of implementing any management decision (James and Teng, 1979).

Three types of loss estimates may be distinguished: prospective, current and retrospective. Prospective losses are predicted by models as losses likely to be incurred given the known historical and present data on disease intensity and/or predicted progress of the epidemic. Current losses are predicted using disease data collected up to the day of prediction and are losses experienced in the system at that time. Retrospective loss predictions are made after part of or the complete epidemic.

Prospective loss estimates are needed to determine the difference between cost of control inputs (like fungicide) and the economic benefits (reduction of loss) of a disease management strategy. These loss estimates would enable rationalisation of chemical use on an objective basis and can be used to enhance systems that predict disease development alone. Prediction of the current level of loss enables decisions on the need for further disease control. For example, in sugarbeet crops, towards the end of the growing season, farmers in Minnesota are often faced with the question of whether they can stop spraying for *Cercospora* leafspot and not incur any economic loss since more than 90% of potential production has been achieved.

Retrospective estimates of crop loss facilitate the allocation of resources for plant pathology research and extension. In this regard, crop loss assessment may be viewed as 'problem definition', providing objective criteria to determine priorities. Crop loss data collected over many seasons further enable the evaluation of changing agricultural technology on disease and its economic effects on crops (James and Teng, 1979).

II. DATA REQUIREMENTS FOR MODEL DEVELOPMENT

Crop loss models need to represent the dynamic interaction between pathogen and host, and the resulting disease that causes loss. James (1974) has suggested that disease–loss models be capable of predicting loss for the range of epidemic types encountered at different crop development stages. Thus, the data used for developing models would have to be collected from several years and locations, and incorporate husbandry representative of the farming system studied. Data for the development of disease–loss models may be collected through either experiments or survey procedures; in both, the objective is to generate 'treatments' representing a wide range of epidemics (James and Teng, 1979).

A. Measurement of Host Variables

The relationship between yield/crop loss, crop development and disease intensity has been conceptualised as a response surface (Teng and Gaunt, 1981), where the amount of loss is dependent on amount of disease and crop development stage. This conceptualisation serves as a convenient means for examining what data are needed to model part or all of the surface.

The same amount of disease at different host development stages may cause different yield losses; this phenomenon was observed as long ago as 1927 (Kirby and Archer, 1927). Host development has commonly been quantified as growth stages using numerals with values that may not have phenological meaning. Taking the cereals as an example, three growth stage keys are the Feekes Scale (Large, 1954), the Romig Scale (Calpouzos *et al.*, 1976) and the decimal code (Zadoks *et al.*, 1974). The same development stage, anthesis, has numberical values of 10.5 (Feekes), 16–18 (Romig) and 60–69 (decimal). This difference presents no problem when one is interested only in a qualitative description of growth stage. However, a quantitative meaning is needed for simulation modelling and response surface modelling, otherwise, so-called weighting factors ('fudge factors') are assigned to indicate the relative importance of growth stages to economic yield. Models have been developed which use growth stage values as an input variable for loss prediction, and one must question the basis for assigning numbers to development (Calpouzos *et al.*, 1976). A significant step in improving the quantification of crop growth stages will have been made if general agreement can be found for a universal code to describe these stages using meaningful criteria. In the USA, a pilot national pest survey and detection system sponsored by the United States Department of Agriculture is researching the development of such a code.

Another host plant variable that is commonly measured for loss modelling is yield, which is considered to be any measurable produce of economic value from a crop (Chiarappa, 1981, p. 123). Yield may be affected by disease directly (as with a foliar disease on a forage crop) or indirectly (as with a foliar disease on a cereal). Because loss is calculated as a difference between two defined yield levels (Chiarappa, 1981; Zadoks and Schein, 1979), it is important that errors be minimised in yield sampling. Relatively little research has been conducted to determine sample sizes needed to detect differences in yield caused by incremental disease intensities. The problem of interplot interference, resulting in data that may misrepresent disease dynamics and yield responses in farmers' fields, is now receiving much attention (Teng and Oshima, 1983). A related problem is the inherent variability in crop growth, resulting in yield samples from small plots having high variance. One popular statistical textbook has cautioned its readers that, if coefficients of variation in corn yield are outside the interval, 5 to 15%, then the validity of the experiment needs to be examined further (Snedecor and Cochran, 1967). Yield estimation remains a major problem in crop loss studies, especially the estimation of potential yield in the absence of disease (Large, 1966).

The prediction of losses caused by low disease intensities is needed in disease management, so that control may be initiated before the crop is destroyed. However, because of the inherent variability in both yield and disease intensity measurements between plants, most data on disease–loss, when graphed as loss versus disease intensity, exhibit heteroscedasticity (uneven variance) at the lower and upper ends of the scatterplot (Teng, 1981a). The ability to quantify low levels of yield loss, caused by low levels of disease, is essential for testing hypotheses on minimum disease thresholds, since some models suggest that no loss is caused below a certain disease intensity (section IV,C).

Green leaf area or undiseased leaf area has been suggested as a predictor of loss (Gaunt, 1980) and yield (Carver and Griffiths, 1983) in cereals. Because disease expresses itself in many ways other than the reduction of photosynthetically active leaf area, green leaf area may be applicable only to those crops where the major proportion of loss is due to one yield component (e.g. 1000-kernel weight). Furthermore, it is questionable whether leaf area has sufficient repeatability among sites and seasons to be an input variable for loss models.

B. Measurement of Disease Variables

Disease, expressed in various forms, is the input variable for the majority of loss models. There are two problems associated with disease measurements from a sample. Firstly, variability in mean disease per plant changes with the stage of the epidemic, although this may be partially minimised by appropriate sample size and method (James and Teng, 1979). Secondly, field methods for disease assessment use the human eye as a 'sensor', introducing a degree of subjectivity and variability between observers. In using linear least-squares regression to quantify the relationship between disease and loss, disease is often considered the independent variable, which is hypothetically without variance. This latter assumption may be satisfied with single tillers or plants, where each tiller or plant is treated as a case for regression and each disease value may be assumed to have zero variance (Richardson *et al.*, 1975), but is never the case with plot experiments, where disease intensity is determined as a mean from *n* plants per plot.

A major consideration in any method of quantifying disease is whether the method captures the concept of disease associated with a certain pathogen. A rust disease, showing discrete lesions on leaves, may be quantified by using 'disease severity' to include actual area covered by lesions, accompanying chlorosis and necrosis (Teng, 1983). It is more difficult to quantify a systemic bacterial infection or wilt into a severity value. Some workers have used grading systems for viral and bacterial diseases (e.g. 0 = healthy and 10 = dead) and the grades as input variables for predicting loss. One must question the statistical appropriateness of using what may be disease 'ranks' for regression analysis. Horsfall and Cowling (1978) use 'disease intensity' to mean 'disease severity' and it is suggested here that their term may be better applied to systemic diseases where no simple estimation of plant area affected can be made.

For a model to depict the field system realistically, a repeatable method of measuring disease, which minimises variability in the biological system and in actual assessment, is needed to collect data for model development. A portable version of a video image analysis instrument should be capable of minimising variance due to measurement method, and significant progress has been made to achieve this (Lindow, 1983). However, visual disease assessment methods will continue in common use for some time yet, and because of this, research into increasing their accuracy is still required. Herbert (1982) noted the apparent weakness of using the Weber–Fechner Law for designing assessment keys such as the Horsfall–Barratt Scale (Horsfall and Barrett, 1945). James (1974) was of the opinion that, with an experienced assessor, it is possible to use a key with linear increments of disease severity, rather than the log relationship suggested by the Horsfall–Barratt Scale. It may well be that visual acuity is related to stimulus (disease symptoms) by a linear or power function, and the lack of definitive data points to the need for more comparative studies on methods of visual assessment.

C. Measurement of Disease Progress

One major difficulty in generating data for disease–loss modelling is to anticipate how many host and disease data to collect. Large (1966) described the need for 'preliminary portfolios' on healthy and diseased plants to aid in interpreting the disease–loss relationship. In general, disease progress curves are required for each treatment plot of a field experiment or a field in a survey. However, in some crops such as the cereals, *a priori* knowledge of yield physiology and disease epidemiology provides a means for optimising data collection in experiments (Teng and Gaunt, 1981). The potential effect of early versus late or long versus short-duration epidemics in reducing cereal yield through effects on certain yield components may be postulated using conceptual models of loss. As a result, some periods of the growing season may be identified for more intensive sampling, to coincide with natural occurrence of the disease and its period of expected impact on crop yield. For many crops, weekly or biweekly disease and host assessments are sufficient to provide the portfolios for describing disease–loss.

Methods for the mathematical description of disease progress curves are described elsewhere (Chapter 2). In disease–loss modelling, 'predictors' or 'independent variables' may be derived from part of or a complete disease progress curve. It would therefore appear important to ensure that progress curves for each treatment or case have minimal variance. Van der Plank (1963) provided a convenient means for linearising progress curves through the logit transformation while Large (1952) has used the probit transformation. There do not appear to be good guidelines as to whether original raw data on disease should be used for model development, or whether such data should be fitted to a non-linear model or linear model before use in loss modelling.

In collecting sequential disease and crop data, the size and type of host unit for assessment requires determination. The potential use of the model is a consideration in what plant part to assess. Common host units in loss studies are single/multiple leaves, tillers (shoots), whole plants, hills and plots. Many disease–loss models are based on the use of a disease assessment from just one leaf, in cereals commonly the flag. Unless there is strong reason for doing this, data collected from more than one plant part will enable more flexible modelling and the development of models to meet several objectives (section III).

III. LOSS PREDICTION MODELS

A. Statistical Procedures

Disease–loss models have been developed from the extensive use of regression and simulation techniques. This led Teng (1981c) to suggest that such models could be categorised as empirical, explanatory and conceptual. Empirical models have been widely applied for loss assessment and are developed using data from field experiments. Simulation models of disease–loss require an explanatory crop model (Loomis *et al.*, 1979) for predicting the yield outcome given disease. Conceptual models of disease–loss use existing knowledge of crop physiology and disease effects on yield to postulate probable responses of yield loss to disease intensity. These probable responses may be represented as generalised curves or three-dimensional response surfaces (Teng and Gaunt, 1981; Gaunt, 1980).

In using linear or non-linear regression for loss modelling, the general procedure is to identify a dependent variable (e.g. yield, yield loss) and quantify the manner in which this changes with an independent variable or variables. The independent variable(s) commonly represents disease intensity expressed in relation to some point in time or as a summation variable such as area under the disease progress curve (AUDPC). Although the variables used for predicting the outcome of a dependent variable are assumed to be independent, this is seldom the case with plant diseases. The use of regression in disease epidemiology and crop loss modelling appears to violate many of the assumptions in the technique, and we generally rationlise away these violations with biological arguments. For example, the problem of autocorrelation in the residuals and variables is a real one and Butt and Royle (1974) noted that plant pathologists often disregard the presence of 'biological autocorrelation'. Further, least-squares regression assumes normality in the characteristic sampled and used for modelling (Draper and Smith, 1966); disease incidence/severity may often not be normal during the early stages of an epidemic.

Regression models of disease–loss have seldom been subjected to independent validation with data not used in their development (Burleigh, 1980). It is often considered that there is an implicit validation process used in accepting a regression model with statistical criteria like r^2, F, t and s (Teng, 1981a). However, the literature shows that most authors have used either one or two of these statistics to evaluate their models, and that often models have been accepted with a high F value but a low r^2 (Teng and Gaunt, 1981). More than one evaluative statistic should be used to accept a model. James and Teng (1979) considered r^2 the single most important indicator of the predictive ability of a model. However, the choice of an evaluative statistic appears to have been a largely subjective process!

In addition to linear regression, non-linear regression has been used for modelling disease–loss (Madden *et al.*, 1981a,b). Multivariate statistical techniques such as principal components analysis and canonical correlation have also recently been used in crop loss studies (Stynes, 1975).

B. Empirical Models

The following model types are considered to be empirical models: single point, multiple point, AUDPC, response surface, synoptic and integral. All are derived from data generated through field experiments or sampling.

1. Single Point Models

At least three types of single point models may be distinguished. The most common predicts loss from disease intensity (severity or incidence) at a specific time in the crop's growth (either a critical development stage or a predetermined number of days into the growing season). With some diseases, workers have been able to identify a 'critical' time during which the crop appears to be most sensitive to disease. For example, the prediction of loss due to wheat stem rust from the model,

$$\% \text{ Loss} = -25.53 + 27.17 \ln(X)$$

where X = percentage stem rust severity at 3/4 berry stage (Romig and Calpouzos, 1970). The model explained 98.6% of variation in yield loss due to rust and suggested the presence of a minimum threshold value of rust (2.6%) below which no loss was caused (Teng and Gaunt, 1981). Development of a critical point model should not be taken to mean that there are no other crop stages which could respond to disease, but rather, that a particular stage shows good statistical correlation with loss (James and Teng, 1979).

Working with potato late blight in Sweden, Olofsson (1968) developed the concept of a 'critical injury threshold' of blight. The period before this critical threshold was considered to be blight free, and loss in tuber yield was predicted from the number of blight-free days; e.g. for south Sweden,

$$Y = 234.0 - 1.706X$$

where Y = yield in tons per ha and X = number of blight-free days (Olofsson, 1968). While Olofsson's (1968) model required detection of a relatively low level of disease, Large (1966) used an analogous concept of a critical injury level causing loss, with much higher blight severities. In Large's (1966) method, 75% late blight was taken to be the critical injury level, with 4% yield loss if this was reached in mid-September, 13% loss, 28% loss and 50% loss, respectively, if the critical injury level was reached in late August, mid-August or late July. The above two critical point models have not found application outside their immediate environments and Zadoks and Schein (1979) have expressed concern about the assumptions on blight epidemiology in such models. In the light of our present understanding of disease–loss, it would appear that these models could be reformulated so that loss is predicted from severity at a known time after planting, as with most single point models.

Single point models are generally suited for predicting loss caused by short duration epidemics, affecting one or two yield components, and their use assumes that epidemics have a relatively stable apparent infection rate. This last point is important; single point models assume that the disease epidemic will develop before and after the 'critical point' in the same manner as the epidemics encountered in experiments.

2. Multiple Point Models

Multiple point models predict loss using disease variables derived from sequential assessments of disease progress during crop growth. The variables are either disease increments during a specified period (James *et al.*, 1972) or actual disease severities at specific growth stages (Burleigh *et al.*, 1972; Teng *et al.*, 1979).

Working on potato late blight, James *et al.* (1972) divided the potato growing season into 9 periods: period 1 for days up to August 5, and subsequent weekly periods, up to period 9 ending September 30. These were denoted, respectively, $X1$ to $X9$, with an additional five variables, $X10$ to $X14$, corresponding to $X5$ to $X9$ for late infections. The blight severity increment for each period was used as the independent variables to predict loss, giving a general model:

$$Y = 1.867X1 + 0.446X2 + 1.144X3 + 0.628X4 + 0.193X5 + 0.180X6 - 0.012X7 + 0.343X8 + 0.829X9 + 0.170X10 + 0.067X11 + 0.044X12 - 0.155X13 - 0.828X14$$

where $Y = \%$ yield loss. The model was further reduced in practice to two simpler forms, one for early infections and one for late infections, as defined by the workers. The original model explained 97.6% of the variation in loss and had a low standard error of estimate of loss (residual mean square of 10.2). The model enabled loss estimation when epidemics were encountered which had variable infection rates during the growing season. The model of James *et al.* (1972) was accepted by its developers using, in addition to statistical criteria, three other subjective criteria which they felt were important: the residual mean square should be low, no negative partial regression coefficient should have a net effect of increasing yield, and no single period should exert an overriding effect on the final loss percentage.

With wheat leaf rust, Burleigh *et al.* (1972) used four growth stages (boot, heading, early berry, early dough) and rust assessments on either flag leaf or the whole tiller as independent variables to develop a multiple point model for loss prediction. The authors found that, although they could identify critical point models from their data, the best model was one which used assessments from three growth stages:

$$Y = 5.3788 + 5.5260X2 - 0.3308X5 + 0.5019X7$$

where $Y = \%$ loss, $X2 = \%$ rust severity per tiller at boot, $X5 = \%$ rust severity on flag leaf at early berry and $X7 = \%$ rust severity on flag leaf at early dough. The model explained 79% of the variation in loss and had a standard error of estimate of loss of 9%.

Both James *et al.* (1972) and Burleigh *et al.* (1972) attempted to evaluate the predictive ability of their models using independent data sets (Fig. 1).

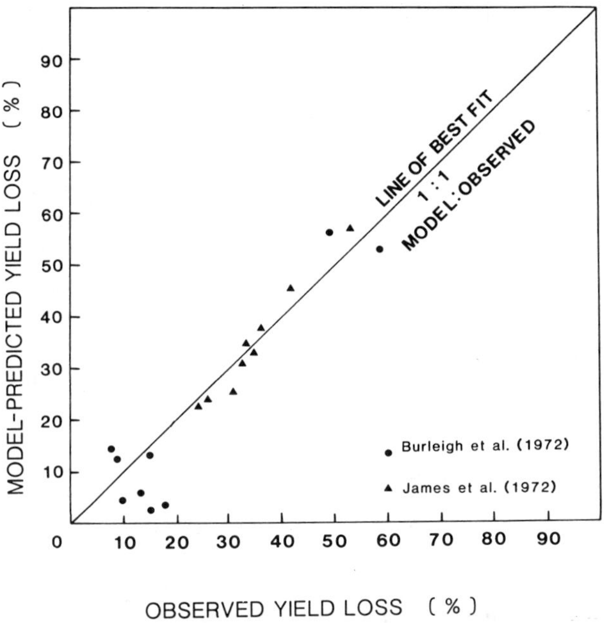

Fig. 1. Validation of model-predicted yield loss against observed yield loss. From data of Burleigh *et al.* (1972) and James *et al.* (1972).

Even though no statistical tests were used to compare model predictions against observed losses given the same disease variables, the scatter of points near the line of gradient = 1 shows that predictions were close to those observed.

3. Integral Models

Integral models predict loss using input variables that are integrals or summations of disease severities over a defined period of crop growth. A common input variable is area under disease progress curve (AUDPC).

Van der Plank (1963) first suggested that yield loss may be related to the AUDPC, and demonstrated a linear relationship using wheat stem rust data of Kingsolver *et al.* (1959). Van der Plank (1963) further cautioned that the ratio between different areas is independent of infection rate, and hence it may be difficult to partition the relative importance of disease at different crop growth stages. A long duration epidemic with low infection rate may have the same area as a short duration epidemic with high infection rate, while the effects of each epidemic on yield may be very different. Generally,

AUDPC models for disease–loss have been for short-duration, late epidemics, e.g.

$$\%\ \text{Yield loss} = 0.43\ \text{AUDPC} - 14.95$$

(Schneider *et al.*, 1976). In this model for *Cercospora* leafspot on cowpea, disease area was summed for severities after flowering. With wheat stem and leaf rust, Buchenau (1975) developed a predictive system for loss using an estimated rate of increase in rust severity coupled with an AUDPC model. The model explained 86% of the variation in yield loss and indicated a 1:1 relationship between loss and area.

An improvement to the AUDPC concept was made by Hills *et al.* (1980), who distinguished between long- and short-duration epidemics of powdery mildew on sugarbeet that had approximately similar areas. The input variable for loss prediction in their model was a weighted average of percentage mature leaf area diseased (% MLAD) assessed at several tiems, i.e.

$$\%\ \text{MLAD} = \sum[W_i(\%\ \text{MLAD})]/\sum W_i$$

where W_i are weeks to October 1 for each assessment date. One of their models was percentage crop loss = 0.238 + 0.636 (% MLAD), which explained 84% of the variation in loss. Hills and Worker (1983) subsequently developed disease thresholds to guide fungicide application for powdery mildew control.

Teng *et al.* (1981) predicted losses due to ozone on alfalfa using an integral form of the multiple point model, where

$$\text{Total dry matter loss, } \% = \sum_{i=1}^{N} Y_i \cdot \mathrm{d}t$$

where Y_i is an equation for each dt describing the relationship between ozone dose and dry matter reduction. In the model, a negative feedback was used (1.0 − dY), in which the remaining proportion of loss was multiplied by the potential loss for a dt, thereby ensuring that only unaffected biomass was counted in the integrated loss value. The authors used this approach in contrast with the multiple point models as the plant part affected by ozone was the economic yield (cf. cereals) where leaves may be diseased but effects are expressed indirectly at the end of the season in grains.

4. Response Surface Models

The relationship between yield loss, disease severity and crop development stage may be conceptualized as a multidimension response surface,

$$\%\ \text{Yield loss} = \mathrm{f}(X, T)$$

where X = disease severity and T = crop development stage (Teng and Gaunt, 1981). A multiple point model may be represented as:

$$\%\ \text{Yield loss} = f(X_1, X_i, \ldots X_T)$$

where X = disease severity at a specific crop stage, i to T. Other epidemic or crop variables have been used to create the response surface, for example, the model of Calpouzos *et al.* (1976) for wheat stem rust estimated loss from

$$\%\ \text{Yield loss} = f(X1, X2)$$

where $X1$ = slope of epidemic line (infection rate) and $X2$ = epidemic onset stage measured using the Romig Scale.

Response surface models require considerably more data to develop than all the models discussed previously. Furthermore, the use of traditional, replicated designs for field experiments may be insufficient to provide the number of treatment points needed to define the surface. Some preliminary work suggests that increased accuracy in loss prediction may be obtained using these models relative to CP (critical point), MP (multiple point) and AUDPC models (Teng and Montgomery, 1982). Conceptually, these models make more realistic use of regression techniques than do traditional replicated designs (Teng and Oshima, 1983).

5. Non-linear Yield Loss Models

Least-square regression is inherently a linear statistical technique, while many biological responses are non-linear. Some workers have transformed the disease variables for predicting loss, implying a non-linear relationship, but used regression to define a linear model. An example is the wheat stem rust model of Romig and Calpouzos (1970),

$$\%\ \text{Yield loss} = -25.53 + 27.17\ \ln(X)$$

and the barley mildew model of Large and Doling (1962),

$$\%\ \text{Yield loss} = 2.5\sqrt{\text{Mildew severity at Feekes 10.5}}$$

In linear regression models, the intercept on the x axis is the disease intensity of zero loss. Above this disease intensity, incremental loss is related to incremental disease by parameters of the equation. There is as yet no consensus on the generalised form of the relationship between disease intensity and yield loss (see section IV,C for detailed discussion). Madden *et al.* (1981a,b) have proposed the use of a density function, the Weibull distribution, for non-linear fits of yield loss to disease severity. The general loss model has the form:

$$L_i = 1 - \exp\{-[(X_i - d)/b]^s\} + u_i$$

where L_i is the proportion of loss for the ith experimental unit, X_i is a univariate disease measurement for the ith experimental unit, u_i is the ith unit of random, unexplained error and d, b and s are parameters. The variable X may be disease intensity at one point in a crop's growth or an integrated measure of disease like AUDPC. Madden *et al.* (1981a) tested this model on potato early blight, and arrived at:

$$L = 1 - \exp[-(\text{final severity} - 0.17)/0.57]^{0.51}$$

and

$$L = 1 - \exp[-(\text{AUDPC} - 5.29)/12.46]^{0.40}$$

In the generalised model, d is the disease threshold below which no loss occurred, b is a slope parameter and s is a shape parameter. The above models had threshold values of 0.17 for the final severity model and 5.29 for the AUDPC model. These workers compared the coefficients of multiple determination for the linear, multiple point model with non-linear models and found no difference when the variable was AUDPC but an improvement from 0.76 to 0.83 when the variable was final disease severity. Furthermore, the non-linear model with final disease severity as the input variable gave a better fit to data than did the linear multiple point model (Madden *et al.*, 1981b). A non-linear model for loss prediction has much biological sense. However, most scattergrams of yield loss (ordinate) plotted against disease intensity (abscissa) display heteroscedasticity at the lower levels of disease and loss (Teng, 1981b). The extreme ends of the yield loss–disease intensity curve are responsible for the non-linear shape (Fig. 1). At low disease intensity and low yield loss, coefficients of variation in disease and yield loss measurements are commonly 5–15%. With this variation at low disease and loss, it may be difficult to quantify a minimum disease threshold conclusively.

6. Synoptic Models

The majority of yield loss models have been developed for single diseases, while in crop production multiple diseases or multiple constraints are more common (James and Teng, 1979). To model the multiple constraint situation adequately would require a large data set in which many levels of each constraint (factor) are represented. In general, researchers have adopted sample surveys to obtain these data. Stynes (1975) introduced the expression 'synoptic' to describe his modelling procedures and this term will be used to describe those models developed for yield or loss estimation using multiple input variables from several diseases or non-disease factors. An early attempt at including multiple input variables for yield estimation was made by Sallans (1948), who related yield of winter wheat to five variables: rainfall during cropping season, rainfall during the preceding autumn and spring, average daily temperature during cropping summer, percent damage due to insects, and percent common root rot. The model explained 78% of the 10-year variation in yield and may be used to estimate yield in the presence/absence of root rot, and from this, loss. Pinstrup-Andersen *et al.* (1976) developed a multiple regression model for predicting edible bean yield using eight variables, of which rust, bacterial blight and angular leaf spot were significant yield depressors. As in the Sallans (1948) model, the Pinstrup-Andersen (1976) model assumed that input variables exerted independent effects on yield, and that correlation between variables had no significant effect on the validity of using regression.

The 'synoptic' approach of Stynes (1975) led to the development of realistic models that predicted yield and explained yield variation from multiple factors in wheat fields in South Australia. The data were grouped into major subsets of climate, soil, past field history, plant variables and pathogen variables. Principal components analysis and canonical correlation were then used on selected sub-sets of the data to identify interactions, as well as to develop new variables where appropriate to minimise correlation between variables. The original and new variables were then used for developing models to explain wheat growth at fixed times during the season, and to predict yield at maturity, from multiple regression analyses. Many predictive models were developed from this study. For example, the model which predicted wheat yield at maturity used 12 input variables and explained 84% of the yield variation with a significant *F*-test (Stynes 1975, 1980).

In a study of determinants of yield in field peas, Wiese (1980) examined some 40 variables and their correlation to final yield, and derived an overall model which explained 82% of the variation in yield using 12 variables, none of which was a disease. When the model was used in subsequent years, it did not predict yield accurately in many fields (M. V. Wiese, personal communication). A possible explanation is that although many variables were intensively sampled during data collection for model development, not enough levels of all variables were encountered to enable a test of the relative contribution of each variable to yield. This is a common problem in using the synoptic approach for research on crop losses.

7. *Comparison and Choice of Models*

Empirical models depend largely on the validity of the data from which they have been developed, and in general do not claim to explain causation. They are 'best-fit' models, although some workers have tried to ensure that biological rules are not violated (Burleigh *et al.*, 1972; Stynes, 1975). Therefore, more than one model form may be expected to fit a data set, presenting the dilemma of choosing between models of the same system and developing some criteria for evaluation. Model development is inextricably linked to the potential use (objective) of the model (Teng, 1981b), and any comparison based on statistical criteria alone is likely to be insufficient.

James (1974) stated that critical point models are particularly suited for estimating losses due to short-duration epidemics with stable infection rates, and which occur relatively late in a crop's growth. He further stated that multiple point models were best used for long-duration epidemics with variable infection rates. Schneider *et al.* (1976) found that, with *Cercospora* leafspot on cowpea, a disease that commonly occurs after anthesis and is of short duration, an AUDPC model explained more of the variation in loss than a critical point model, although the latter was satisfactory. When comparing linear models (single point, AUDPC and multiple point), although statistics such as the coefficient of multiple determination, F, s and t may be used, the biological logic behind such models also needs to be considered. This is important if the empirical modelling process is not to be mere curve fitting. On theoretical grounds, it would be expected that a model for loss prediction based on disease assessments at several growth stages would be more reliable than one based on a single assessment, since the dynamics of the epidemic would be better represented. There is much evidence to show that this is true for long-duration epidemics (Kirby and Archer, 1927; James *et al.*, 1972; Burleigh *et al.*, 1972; Teng *et al.*, 1979), leading James and Teng (1979) to suggest that the precision of a critical point model can almost always be improved by a multiple point model developed from the same data. However, this appears to be true only when comparing linear models. Madden *et al.* (1981b), using data on peanut *Cercospora* leafspot and potato early blight, compared linear multiple point models with two non-linear Weibull models, one using AUDPC and the other final disease severity as input variables. Both non-linear models were judged better predictors of loss than the linear multiple point model, in each disease, leading Madden *et al.* (1981b) to conclude that multiple point models do not always offer increased precision over single point or AUDPC models, as suggested by James and Teng (1979).

Epidemics may be represented by disease progress curves which have the same area, but differ in disease onset time, infection rate, duration and final severity. A long-duration epidemic with early onset, low infection rate and low final severity has quite a different effect on crop yield when compared with a short duration epidemic with late onset, high infection rate and high final severity (Teng and Gaunt, 1981). While improvements in model fit to data will be obtained using a non-linear form relative to linear, the ability of a model to predict yield losses will need to be tested using data sets from other seasons. It may well be that a valid comparison of multiple point versus single point or AUDPC models may be made using the same general non-linear model, and not by comparing a non-linear single point or AUDPC model with a linear multiple point model. Furthermore, there is still uncertainty as to the form of the disease–loss response and its change as a plant or crop matures (section IV,C).

The choice of a model for loss prediction is largely influenced by the objective. For large-area crop loss assessment, in which many fields are to be visited, single point models have been favoured because they require one disease assessment (Madden, 1983). AUDPC and multiple point models have been preferred in loss predictions using loss models coupled to epidemic simulation models (Teng *et al.*, 1978, 1980). With increasing activity in pest management scouting in many parts of the USA, disease intensity data have become available for many crops on a weekly basis, data which are well suited for loss prediction through single point, AUDPC or multiple point models (Teng, 1983). These data may also be used to develop synoptic models for crop losses, when collected using acceptable procedures. Currently, in the USA, systematic collection of pest and crop data from representative fields is being encouraged by two national systems, both sponsored by the United States Department of Agriculture.

C. System Models

The apparent inability of empirical models to have explanatory value has led researchers to suggest using crop simulation models as a means of yield and loss prediction (Ruesink, 1981; Teng, 1981c). Crop models simulate growth and development at the physiological process level (Loomis *et al.*, 1979) and at the tissue level (e.g. Sands *et al.*, 1979). With the process level models, functional relationships are defined for interactions between pathogen and host, their interference with basic processes such as respiration, photosynthesis and evapotranspiration, and the net effect on plant growth and development (Loomis and Adams, 1980; Teng, 1981c). These models allow an understanding of the physiology of pathogen–host interactions; they have research value but are often too cumbersome for practical loss prediction. Process-level models have been developed for major crops such as soybean, corn, wheat, potato and alfalfa.

The second group of crop models may be exemplified by the potato models developed in Australia for yield estimation in fields (Sands *et al.*, 1979; Hackett *et al.*, 1979) and the large-area crop production models of Holt and co-workers at Purdue University, Indiana (D. Holt, personal communication; Teng and Oshima, 1983). The potato model predicts tuber bulking from the growth and development rates of several plant parts, as influenced by temperature, insolation, day length and soil–water deficit (Sands *et al.*, 1979). The Purdue model may be viewed as an elaborate county-level model for yield prediction in which a potential maximum yield is reduced by multipliers either calculated from sub-system models or estimated.

In using crop models for loss prediction, a reference yield is calculated for set conditions of zero disease, and the model is then run under the same conditions but with disease present. The dynamics of disease progression may need to be modelled by an epidemic model (empirical or simulation) for realistic representation of field conditions. Although the possibilities exist for the above, there has been no reported loss prediction for any crop region using this approach. As Teng and Oshima (1983) have noted, there are insufficient data available to relate pathogen growth to crop growth and development. Part of the reason is the lack of a 'systems approach' in identifying research needs that further our understanding of yield and crop loss.

IV. SOME CURRENT ISSUES IN LOSS MODELLING

A. Predictive Ability of a Model

Mathematical models provide a convenient means of synthesising voluminous data into a usable form. However a model is defined, the question remains as to how valid any model is in representing the biological system, and in predicting the behaviour of that system. The evaluation of models is receiving increased attention from modellers (Teng, 1981b), especially in relation to verification and validation procedures. Verification is concerned with how well a model has represented current hypotheses and knowledge of a system; for example, theoretical considerations on the shape of the disease–loss relationship will affect verification of an empirical model. Validation is concerned with establishing similarities between real-world observations and predictions made by a model, when subject to the same input variables. Few disease–loss models have received validation of their predictive ability, especially with data not previously used for developing the model. The models of Burleigh *et al.* (1972) and James *et al.* (1972) appear to predict losses accurately (Fig. 1) on diseased plots. However, no effort seems to have been made to test the predictive ability of models against regional losses. Empirical and systems models for loss prediction are commonly developed from plot data. Yet, regional loss predictions are generally made on production units bigger than a plot, using these plot-level models. Perhaps all regional loss estimates that have been published suffer from the representational error of Van der Plank (1963), and, perhaps because of this, there are published reports of losses that exceed 100% when losses from individual pests are totalled for the same crop (James, 1980).

With empirical regression models, statistics used for accepting the model may give an indication of the predictive ability of the model. In this regard, the coefficient of multiple determination, which shows the proportion of variation in loss explained by the input variables of the model, and the *s* statistic, the standard error of estimate of loss predicted by the model, are two important statistics of several commonly used for evaluating regression models (Teng, 1981a). These statistics do show, however, only the predictive ability of the model for the data set used in model development; if conditions other than those in the model data set are encountered, spurious losses will be predicted from the model. This is particularly important with single point models, which depend entirely on the similarity in disease progress between model data set and the new environment for prediction. There is thus an important requirement for rigorous, independent testing of the predictive ability of empirical models, whether they are linear or non-linear.

It has been suggested previously that disease–loss models may be used in disease management to rationalise fungicide use based on cost–benefit criteria provided by these models (Teng *et al.*, 1978; James and Teng, 1979). The modus operandi would include a prediction of loss through empirical models using projected disease progress, followed by calculation of the monetary value of the loss from a known potential yield, and then comparison with control cost. If no spray is applied, then a predicted monetary loss would be incurred. However, it does not follow that, if a spray or spraying regime is adopted, extra revenue will be generated. Our experience with empirical loss models as decision aids to guide spraying has shown that the logic in yield loss prediction does not agree with the need to forecast gain (profit or recoverable loss) since the experimental data used to develop loss models generally do not allow prediction of gain [i.e gain is not (1.0 − loss)]. In one such decision aid (Teng and Montgomery, 1982), it was found that growers were more interested in a positive prognosis (how much extra gain from spraying or how much recoverable loss) than in the negative prognosis predicted by the loss models (how much loss if no control is practised). We need to re-think the collection of experimental data that would enable not only loss but gain prediction!

The prediction of losses that have not yet occurred is commonly done in conjunction with a method of disease prediction. Linear prediction of disease progress has been attempted using non-linear axes for disease (Buchenau, 1970; Calpouzos *et al.*, 1976). Simulation models have also been examined for predicting disease progress (James, 1974; Teng *et al.*, 1978). In either case, predicted severities are used as input for loss models. Predicting future losses for decision making in crop protection should, on theoretical grounds, be highly desirable. Yet there is little evidence of this being used in practice, due partly to the uncertainties associated with any predictions.

B. Experimentation for Loss Modelling

Field experiments to generate data for loss modelling are either paired treatment or multiple treatment (LeClerg, 1971). The objective is to have treatments that would result in a representative range of epidemics, through use of cultivars, location, fungicides and husbandry. Van der Plank (1963) warned of the 'cryptic error' in all experiments which have contiguous plots with epidemics of different severities, and how this may result in a 'representational error' when applying results from plot experiments to farmers' fields. James *et al.* (1973) quantified this representational error as interplot interference in multiple-treatment experiments; more recently, the phenomenon has been studied in paired plots (Bowen and Teng, 1982). The existence of interplot interference in crop loss experiments creates serious doubts as to the validity and usefulness of all empirical models, in particular the single point models which, as Madden (1983) has found, are the most prevalent model type in the literature. A survey of the literature further shows that most researchers have ignored this phenomenon when conducting loss experiments. There has been recent research into ways of quantifying and correcting for interference (Jenkyn *et al.*, 1979; Bowen and Teng, 1982). It may well be that until we can remove the 'cryptic error; from each plot in a loss experiment, it will be difficult to improve the accuracy of empirical models.

Almost without exception, crop loss experiments have used standard field-plot designs with replication. Modellers are faced with the dilemma of whether to use the mean of each treatment as an observation or to use individual plot data as observations, thereby ignoring the replication. The common use of regression analyses in empirical modelling leads one to question the appropriateness of either approach (Teng and Oshima, 1983). Classical ANOVA (analysis of variance) is designed to demonstrate differences between treatments at defined probability levels (Johnson and Berger, 1982) while, with empirical loss modelling, a response is to be estimated. Significance of the response implies a statistical relationship, but does not prove that any two levels of the response are significantly different from each other. The estimation of yield responses to disease or other factors may best be estimated using response surface methodology (Heady and Dillon, 1961; Dillon, 1968), which emphasises treatments over replications, to enable a thorough definition of the relationship. For example, the three-dimensional response surface relating yield loss to crop growth stage and disease severity (Teng and Gaunt, 1981) would require substantial field experimentation to model, and would lead to trade-off decisions on number of treatments versus number of replications. Response surface methodology would suggest an experimental design with a large number of treatments and *no* replication, so that as much of the surface is explored as is feasible. Furthermore, this lack of replication implies that each case (observation) used for regression is deterministic (i.e. zero variance), thus conforming to one assumption in the technique. It is surprising that this approach has not been tested earlier on field plots, since single plant/tiller approaches for loss modelling (Hau *et al.*, 1980; Richardson *et al.*, 1975; James and Teng, 1979) resemble the zero replication response surface method (see section II,B).

C. Physiological Basis for the Disease–Loss Curve

Although researchers do not generally assign interpretative or causal value to the shape of a disease–loss relationship, it would be reasonable to ensure that such curves and the models describing them are not biologically untenable. Models are, after all, only 'sections of reality' (Kranz and Hau, 1980) and often reflect the researcher's hypothesis regarding the system. In empirical disease–loss modelling, one fundamental question is whether to treat the relationship between disease and loss as a linear or non-linear one. The majority of single disease–yield loss empirical models have assumed a linear relationship between disease and loss, even though researchers intuitively recognise that most biological responses are non-linear. However, the problem of variance in both disease and yield (or loss) at low values makes it difficult to ascertain if apparent deviations from non-linearity are real.

It is generally accepted that part of the response of yield to disease is linear (Fig. 2). Fick (1985), using some concepts from Bardner and Fletcher (1974), suggested that physiologically a plant may respond to increasing pest density in one of the following ways: enhancement, compensation, linear decrease, pest competition and a lower threshold (Fig. 2). Significant loss would therefore occur only when the defensive and compensatory mechanisms of

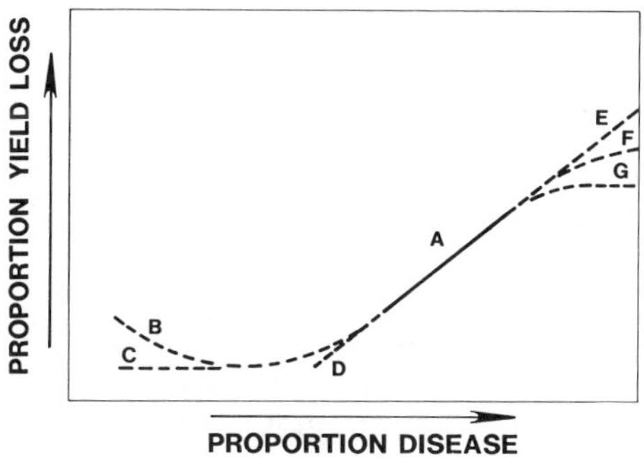

Fig. 2. A conceptual view of the relationship between disease and yield loss. A, D, E, proportionate loss; B, enhancement; C, minimum threshold (compensation); F, pathogen competition; G, maximum threshold.

the plant are overcome. In comparing Fig. 2 with linear regression models, it can be seen that such models omit the tail ends of the curve. A single point model of the form $Y = a + bX$, where $Y = \%$ loss and $X = \%$ disease, may well describe the linear portion of the curve, especially if the data used for model development do not include values outside the linear portion, i.e. very low or very high disease. The presence of a negative value for the intercept (a) would also imply that there is a disease intensity below which no loss is caused; this disease intensity may be calculated by substituting values in the original equation. Conversely, a positive intercept implies that there is loss even without disease! In trying to reconcile yield physiology with statistics, the multiple point model of Burleigh *et al.* (1972) has a negative partial regression coefficient, suggesting that, at the growth stage concerned, disease stimulated yield. Gaunt (1980) has noted that this may not be such a far-fetched hypothesis and Fick (1985) has supported the concept of yield enhancement at low pest densities.

There is insufficient evidence to support the upper disease threshold (Fig. 2) of maximum loss, although indirectly some pathologists have defined a maximum disease severity, such as 37% in the modified Cobb scale (Peterson *et al.*, 1948). There is evidence in the literature to show that, as pest density increases, the loss incurred by each individual decreases due to pest competition, resulting in a departure from a proportionate, linear relationship between pest density and yield (Justesen and Tammes, 1960). In a subsequent paper, Tammes (1961) proposed a generalised curve relating yield to the level of any injurious factor, in which an upper threshold (compensation) level implied no loss beyond that, a middle linear portion implied proportionate loss with disease, and a lower threshold level implied maximum loss. The Tammes (1961) relationship has been used to justify the development of non-linear models to give improved fit of data (Madden *et al.*, 1981b), yet it is incomplete relative to the Fick (1983) representation in that it does not recognise enhancement and pest competition (Fig. 2). It is interesting that while Justesen and Tammes (1960) noted the possibility of an asymptotic behaviour at both ends of the pest density/yield curve, Tammes (1961) in his generalised injury/yield curve set lower and upper thresholds.

Our conceptual view of the shape of the disease–loss curve has a direct influence on the choice of statistical models to represent that curve. Linear regression models appear to be valid for a large portion of the disease-loss relationship, especially at disease intensities above the detection threshold (for most diseases). Linear models may be more meaningful if, during the regression process, there is not an attempt to force the fitted line through the origin, thereby recognising the existence of a lower threshold of disease. Heteroscedasticity of variable points at the lower ends of the disease–loss scattergram often precludes any visual recognition of this threshold (Teng,

1981a). Further, at low disease values, there is much spatial variance in the diseased population and this has allowed researchers to quantify a relationship between disease incidence and disease severity (James and Shih, 1973; Rouse *et al.*, 1981; Teng, 1978). The existence of this relationship suggests that it would be difficult to determine the disease–loss relationship at low levels of disease; the problem is compounded by the inherent variability in yield and loss even in the absence of disease.

Non-linear models have been shown to give improved fit to disease–loss data, in particular, the use of the flexible Weibull distribution function (Madden *et al.*, 1981a). In comparing linear models with non-linear models on the same data set, these workers found higher coefficients of multiple determination with the non-linear models relative to linear (Madden *et al.*, 1981b). The result is not surprising when one considers that empirical modelling resembles a best-fit process, and adopting a linear approach is restrictive in fitting a model to the data (cf. the use of quadratic and cubic functions to improve model fit). Both linear and non-linear models deserve further comparative studies, using experiments which attempt to reduce the variance in dependent and independent variables at low levels of both. This issue is all the more important as often management decisions are made at low levels of disease (for example, the action threshold of Zadoks and Schein, 1979). It may be that we are over-optimistic in trying to use crop loss predictions to identify action thresholds.

There has been much research into nematode population densities and crop responses, and the development of models that recognise yield physiology as well as nematode population dynamics (Seinhorst, 1965; Oostenbrink, 1966; Barker and Olthof, 1976; Ferris, 1978). The model of Seinhorst (1965) is as follows:

$$Y = m + (1 - m)Z^{P-T}$$

where Y is the ratio of yield at nematode density P to yield in absence of nematodes, Z is a constant with value <1, M is the minimum value of the yield ratio when all available space is occupied by nematodes, T is the lower threshold of nematode density, below which no loss occurs. This model, as with those of Justesen and Tammes (1960), is based on the premise that, as nematode density increases, the contribution of each nematode to yield reduction decreases, thereby resulting in a curvilinear effect. There does not appear to be a general consensus on the shape of the nematode density–yield (loss) curve, and Oostenbrink (1971) has presented some elegent arguments for treating the relationship as a linear one. As with other pathogens, nematode–loss models have not achieved wide use in disease management (Barker and Olthof, 1976).

The economic yield of many crops comprises several components, each of which may be affected by disease at different developmental stages of the crop. In cereals, the components commonly recognised are tillers/unit area, tillers/plant, number of spikelets/ear, number of grains/spikelet and 1000-grain weight (Teng and Gaunt, 1981). While most disease–loss models do not recognise that loss may be partitioned into its biological components, Madden (1983) has suggested that the technique of path analysis be used to increase the interpretative value of models. Teng and Gaunt (1981) proposed a conceptual model based on yield component analysis which could serve as a guide for examining the cause–effect relationships implied by models, and for indicating where treatments should be placed in disease–loss experiments, using *a priori* knowledge on the potential effect of constraints on yield components. It may well be that improvements in loss prediction can be achieved by developing models of individual yield components and integrating them into a single loss model. This is an area of research that does not appear to have been initiated at all.

D. Regional Loss Estimates

Earlier, the problem of 'representational error' was discussed as influencing the validity of crop loss estimates made for a big spatial unit (e.g. a county) when models are used that were developed from data generated from plot-sized experimental units. This error is only one of many difficulties encountered in the prediction of historical and future losses for any cropping region. A typical scenario for the assessment of crop losses over a region was depicted by James and Teng (1979) as an organogram, where survey data on pest intensity were coupled with predictive loss models to give field estimates. The cereal surveys of England and Wales are an example of this (King, 1980). In our system in Minnesota, USA, we encountered difficulty in defining the reference yield level for loss assessment per county of the state, since the only yield data available to us, for most crops, were those collected by the State Department of Agriculture, and were for yields in the presence of disease and other production constraints (Teng *et al.*, 1981, 1982). The loss predictive models that were used in Minnesota gave percentage loss estimates for each factor relative to reference yields of zero loss, or no constraint. Therefore, although a county average loss percentage could be calculated, the loss in weight units was determined as

$$\text{Loss (tons or other units)} = [\text{county yield}/(100 - \%\ \text{loss})] - \text{county yield}$$

the rationale being that average county yield is yield constrained by (100 − % loss), and in the absence of this, there will be a linear response to a reference yield of [county yield/(100 − % loss)]. Although the method exemplified above is simplistic, it is often the only one available to those interested in regional loss assessment.

Few crop surveys measure yield/field because of the logistic problems of coordinating collection of such data. In the USA, where there is a thriving pest management industry, such data are often available, although they may be a biased sample. We have attempted to get yield/field data directly from growers after harvest, although the accuracy of such data requires corroboration. For example, Wiese (1980) has found a correlation of only 0.61 between grower estimates of pea yield and research yield samples from the same fields. If reliable yield/field estimates are available for a region, then actual loss/field may be calculated from the potential yield/field, and from these the average actual loss for the region.

The foregoing discussion has pointed to the need for peripheral data that are required to assess regional losses using predictive models. It is illogical to consider loss caused by one factor in isolation from loss caused by other factors. Furthermore, environment and crop husbandry have direct effects on the individual and interactive losses occurring in any crop. Yet few interactive models for loss have been developed that are applicable for regional loss assessment. King's (1982) work on potatoes is exemplary of the type of experimentation required for interactive loss modelling. However it also shows the large resources needed for such research. The technique of spatial aggregation, i.e. identifying cropping units with common attributes that can make them collectively a sampling stratum for loss assessment, only partially overcomes the present lack of interactive models for loss prediction. Stratification is essential for large-area crop production forecasting (Hackett *et al.*, 1979) using models specifically developed for the purpose. The experience gained from these county-level crop models suggests that, with regional loss assessment, we should examine the relevance of the plot-level, single disease model approach that is so prevalent. Perhaps county-level predictive models should be developed for regional crop loss assessment, and perhaps we should concede that yield loss predictive models have no validity for crop loss assessment! (See MacKenzie, 1983, for discussion on crop loss versus yield loss.)

V. CONCLUDING REMARKS

Models are only as reliable as the experimental data used in their development, and crop loss researchers need to be continually aware of this. Because model development is a continuing process where, at any one time, the model represents the 'state-of-the-art' perception of a system, we must actively search for ways to improve the biological and statistical properties of models.

Certainly in the relatively short history of quantitative loss assessment, we have seen the progression from loss assessment tables, such as those of Kirby and Archer (1927), to simple linear equations, to multiple point models, and lastly to non-linear and simulation models. Many statistical techniques and approaches which have the potential to improve the realism and accuracy of crop loss models have not been tested, and there is an urgent need for such testing. However, we must, at the same time, invest effort in improving the quality of the biological data required for modelling. In this chapter, some of the weaknesses associated with biological data on pathogen, host and disease were identified. Data quality remains a constraint on improving the predicitive ability of models.

For the future, improvements in computer technology are likely to affect crop loss modelling. The trend is for both hardware and software to become more user friendly, making quantitative tools more accessible to a larger group of plant pathologists. We can envisage computers helping to refine crop loss models and increasing the use of such models for disease management.

REFERENCES

Bardner, R. and Fletcher, K. E. (1974). *Bull. Entomol. Res.* **64**, 141–160.

Barker, K. R. and Olthof, T. H. A. (1976). *Annu. Rev. Phytopathol.* **14**, 327–353.

Bowen, K. L. and Teng, P. S. (1982). *Phytopathology* **72**, 1134.

Buchenau, G. W. (1970). *S. D. Farm Home Res.* **21**, 31–34.

Buchenau, G. W. (1975). *Phytopathology* **65**, 1317–1318.

Burleigh, J. R. (1980). *In* 'Crop Loss Assessment,' pp. 50–62. *Misc. Publ. Univ. Minn. Agric. Exp. Stn.* **7**.

Burleigh, J. R., Roelfs, A. P. and Eversmeyer, M. G. (1972). *Phytopathology* **62**, 944–946.

Butt, D. J. and Royle, D. J. (1974). *In* 'Epidemics of Plant Diseases' (J. Kranz, ed.), pp. 78–114. Springer-Verlag, Berlin and New York.

Calpouzos, L., Roelfs, A. P., Madson, M. E., Welsh, J. R. and Wilcoxson, R. D. (1976). *Minn. Agric. Exp. Stn. Tech. Bull.* **307**.

Carver, T. L. and Griffiths, E. (1982). *Ann. Appl. Biol.* **101**, 561–572.

Chester, K. S. (1950). *Plant Dis. Rep. Suppl.* **193**, 189–362.

Chiarappa, L. (ed.) (1971). 'Crop Loss Assessment Methods.' Commonwealth Agricultural Bureaux/Food and Agricultural Organization. Farnham Royal, UK.

Chiarappa, L. (ed.) (1981). 'Crop Loss Assessment Methods,' Suppl. 3. Commonwealth Agricultural Bureaux/Food and Agricultural Organization. Farnham Royal, UK.

Dillon, J. L. (1968). 'The Analysis of Response in Crop and Livestock Production.' Pergamon, Oxford.

Draper, N. R. and Smith, H. (1966). 'Applied Regression Analysis.' Wiley, New York.

Ferris, H. (1978). *J. Nematol.* **10**, 341–350.

Fick, G. W. (1985). *In* 'Insect Pest Management Modeling' (C. A. Shoemaker and W. G. Ruesink, eds.). Wiley, New York (in press).

Gaunt, R. E. (1980). *In* 'Crop Loss Assessment,' pp. 98–111. *Misc. Publ. Univ. Minn. Agric. Exp. Stn.* **7**.

Hackett, C., Sands, P. J. and Nix, H. A. (1979). *Field Crops Res.* **2**, 333–347.

Hau, B., Kranz, J., Dengel, H. J. and Hamelink, J. (1980). *In* 'Crop Loss Assessment,' pp. 254–261. *Misc. Publ. Univ. Minn. Agric. Exp. Stn.* **7**.

Heady, E. O. and Dillon, J. L. (1961). 'Agricultural Production Functions.' Iowa State Univ. Press, Ames.

Herbert, T. T. (1982). *Phytopathology* **72**, 1269.

Hills, F. J. and Worker, G. F. (1983). *Plant Dis.* **67**, 654–656.

Hills, F. J., Chiarappa, L. and Geng, S. (1980). *Phytopathology* **70**, 680–682.

Horsfall, J. G. and Barratt, R. W. (1945). *Phytopathology* **35**, 655.

Horsfall, J. G. and Cowling, E. B. (1978). *In* 'Plant Disease' (J. G. Horsfall and E. B. Cowling, eds.), Vol. 3, pp. 119–136. Academic Press, New York.

James, W. C. (1974). *Annu. Rev. Phytopathol.* **12**, 27–48.

James, W. C. (1980). *In* 'Crop Loss Assessment,' pp. 10–16. *Misc. Publ. Univ. Minn., Agric. Exp. Stn.* **7**.

James, W. C. and Shih, C. S. (1973). *Phytopathology* **63**, 183–187.

James, W. C. and Teng, P. S. (1979). *In* 'Applied Biology' (T. H. Coaker, ed.), Vol. 4, pp. 201–267. Academic Press, New York.

James, W. C., Shih, C. S., Hodgson, W. A. and Callbeck, L. C. (1972). *Phytopathology* **62**, 92–96.

James, W. C., Shih, C. S., Callbeck, L. C. and Hodgson, W. A. (1973). *Phytopathology* **63**, 1269–1275.

Jenkyn, J. F., Bainbridge, A., Dyke, G. V. and Todd, A. D. (1979). *Ann. Appl. Biol.* **92**, 11–28.

Johnson, S. B. and Berger, R. D. (1982). *Phytopathology* **72**, 1014–1015.

Jones, A. L. (1983). *In* 'Challenging Problems in Plant Health' (T. Kommedahl and P. H. Williams, eds.), pp. 362–367. Am. Phytopathol. Soc., St. Paul, Minnesota.

Justesen, S. H. and Tammes, P. M. L. (1960). *Tijdschr. Plantenziekten* **66**, 281–287.

King, E. D. (1982). The effects of late blight, Colorado potato beetle and weed populations on yield loss of potato. Ph.D. Thesis, The Pennsylvania State Univ., College Park.

King, J. E. (1980). *In* 'Crop Loss Assessment,' pp. 124–133. *Misc. Publ. Univ. Minn. Agric. Exp. Stn.* **7**.

Kingsolver, C. H., Schmitt, C. G., Peet, C. E. and Bromfield, K. R. (1959). *Plant Dis. Reporter* **43**, 855–862.

Kirby, R. S. and Archer, W. A. (1927). *Plant Dis. Rep. Suppl.* **53**, 110–208.

Kranz, J. and Hau, B. (1980). *Annu. Rev. Phytopathol.* **18**, 67–83.

Large, E. C. (1952). *Plant Pathol.* **1**, 109–117.

Large, E. C. (1954). *Plant Pathol.* **3**, 128–129.

Large, E. C. (1966). *Annu. Rev. Phytopathol.* **4**, 9–28.

Large, E. C. and Doling, D. A. (1962). *Plant Pathol.* **11**, 47–57.

LeClerg, E. L. (1971). *In* 'Crop Loss Assessment Methods' (L. Chiarappa, ed.), pp. 2.1/1–11. Commonwealth Agricultural Bureaux/Food and Agricultural Organization. Farnham Royal U.K.

Lindow, S. E. (1983). *Phytopathology* **73**, 1576–1581.

Loomis, R. S. and Adams, S. S. (1980). *In* 'Crop Loss Assessment,' pp. 112–117. *Misc. Publ. Univ. Minn. Agric. Exp. Stn.* **7**.

Loomis, R. S., Rabbinge, R. and Ng, E. (1979). *Annu. Rev. Plant Physiol.* **30**, 339–367.

MacKenzie, D. R. (1983). *In* 'Challenging Problems in Plant Health' (T. Kommedahl and P. H. Williams, eds.), pp. 82–92. Am. Phytopathol. Soc., St. Paul, Minnesota.

Madden, L. V. (1983). *Phytopathology* **73**, 1591–1596.

Madden, L. V., Pennypacker, S. P. and Kingsolver, C. H. (1981a). *Phytopathol. Z.* **101**, 196–201.

Madden, L. V., Pennypacker, S. P., Antle, C. E. and Kingsolver, S. P. (1981b). *Phytopathology* **71**, 685–689.

Main, C. E. (1983). *In* 'Challenging Problems in Plant Health' (T. Kommedahl and P. H. Williams, eds.), pp. 61–68. Am. Phytopathol. Soc., St. Paul, Minnesota.

Olofsson, B. (1968). *Statens Vaextskyddsanst. Medd.* **14**, 81–93.

Oostenbrink, M. (1966). *Meded. Landbouwhogesch. Wageningen* **66**, 1–46.

Oostenbrink, M. (1971). *Indian J. Nematol.* **1**, 68–74.

Peterson, R. F., Campbell, A. B. and Hannah, A. E. (1948). *Can. J. Res. Sect. C* **26**, 496–500.

Pinstrup-Andersen, P., de Londono, N. and Infante, M. (1976). *PANS* **22**, 359–365.

Richardson, M. J., Jacks, M. and Smith, S. (1975). *Plant Pathol.* **24**, 21–26.

Romig, R. W. and Calpouzos, L. (1970). *Phytopathology* **60**, 1801–1805.

Rouse, D. I., MacKenzie, D. R., Nelson, R. R. and Elliott, V. G. (1981). *Phytopathology* **71**, 1015–1020.

Ruesink, W. G. (1981). *In* 'Plant Protection: Fundamental Aspects' (T. Kommedahl, ed.), pp. 131–134. Burgess, Minneapolis, Minnesota.

Sallans, B. J. (1948). *Sci. Agric. (Ottawa)* **28**, 6–20.

Sands, P. J., Hackett, C. and Nix, H. A.. (1979). *Field Crops Res.* **2**, 309–331.

Schneider, R. W., Williams, R. J. and Sinclair, J. B. (1976). *Phytopathology* **66**, 384–388.

Seinhorst, J. W. (1965). *Nematologica* **11**, 137–154.

Snedecor, G. W. and Cochran, W. G. (1967). 'Statistical Methods.' Iowa State Univ. Press, Ames.

Stynes, B. A. (1975). A synoptic study of wheat. Ph.D. Thesis, Univ. of Adelaide, South Australia.

Stynes, B. A. (1980).). *In* 'Crop Loss Assessment,' pp. 166–175. *Misc. Publ. Univ. Minn. Agric. Exp. Stn.* **7**.

Tammes, P. M. L. (1961). *Tijdschr. Plantenziekten* **67**, 257–263.

Teng, P. S. (1978). System modelling in plant disease management. Ph.D. Thesis, Lincoln College, Univ. of Canterbury, New Zealand.

Teng, P. S. (1981a). *In* 'Crop Loss Assessment Methods' (L. Chiarappa, ed.), Suppl. 3, pp. 51–55. Commonwealth Agricultural Bureaux/Food and Agricultural Organization. Farnham Royal, U.K.

Teng, P. S. (1981b). *Z. Pflanzenkr. Pflanzenschutz* **88**, 49–63.

Teng, P. S. (1981c). *In* 'Plant Protection: Fundamental Aspects' (T. Kommedahl, ed.), pp. 122–127. Burgess, Minneapolis, Minnesota.

Teng, P. S. (1983). *Phytopathology* **73**, 1587–1590.

Teng, P. S. and Gaunt, R. E. (1981). *Agric. Syst.* **6**, 131–154.

Teng, P. S. and Montgomery, P. R. (1982). *Phytopathology* **72**, 1140.

Teng, P. S. and Oshima, R. J. (1983). *In* 'Challenging Problems in Plant Health' (T. Kommedahl and P. H. Williams, eds.), pp. 69–81. Am. Phytopathol. Soc., St. Paul, Minnesota.

Teng, P. S., Blackie, M. J. and Close, R. C. (1978). *Outlook Agric.* **9**, 273–277.

Teng, P. S., Close, R. C. and Blackie, M. J. (1979). *Phytopathology* **69**, 1239–1244.

Teng, P. S., Blackie, M. J. and Close, R. C. (1980). *Agric. Syst.* **5**, 85–104.

Teng, P. S., Krupa, S. V. and Kromroy, K. R. (1981). *Phytopathology* **71**, 908.

Teng, P. S., Stienstra, W. C., Bowden, R. L. and Randall, G. W. (1982). *Phytopathology* **72**, 984.

Van der Plank, J. E. (1963). 'Plant Diseases: Epidemics and Control.' Academic Press, New York.

Wiese, M. V. (1980). *In* 'Crop Loss Assessment,' pp. 262–269. *Misc. Publ. Univ. Minn. Agric. Exp. Stn.* **7**.

Zadoks, J. C. (1972). *Annu. Rev. Phytopathol.* **10**, 253–276.

Zadoks, J. C. and Schein, R. D. (1979). 'Epidemiology and Plant Disease Management.' Oxford Univ. Press, London and New York.

Zadoks, J. C., Chang, T. T. and Konzak, C. F. (1974). *Weed Res.* **14**, 415–421.

9

THE PRACTICAL IMPLEMENTATION OF MODELS IN CROP DISEASE MANAGEMENT

D. J. Butt* and M. J. Jeger†

**Plant Pathology Department, East Malling Research Station, Maidstone, England*

†Department of Plant Pathology and Microbiology, Texas A & M University, College Station, Texas, USA

I. INTRODUCTION

Implementation of a mathematical model means simply the active use of the model for a given purpose. A model of spore transport is implemented if used in an experimental programme. If subsequently simplified to give, for example, the required isolation distance from an inoculum source, it can be implemented in management to decide where a new orchard should be planted. An economic model is implemented if used to determine, *a posteriori*, what would have been the optimal level of chemical control for a given host–pathogen combination, but this is quite different from implementing a decision model in real time on the farm. Implementation of a model demands the definition of a user. In this chapter we are concerned ultimately with the practical implementation of models on the farm, whether the user of the model is a grower, manager, co-operative, adviser or consultant. It would be too restrictive, however, to limit discussion to examples of models actually being implemented on the farm. This chapter would be slender indeed. We shall take a more comprehensive view of implementation, and cover the whole process following the construction of a model by research, through the various stages necessary for a model to make any impact on agricultural practice. In this process the model will likely undergo many changes and may, indeed, play only a minor role in the entity that is finally implemented on the farm.

It would be quite impossible to cover the whole range of crops grown for food, fibre and other industrial uses. We have decided, in the interests of cohesiveness and of dealing with sufficient case studies, to restrict most of our discussion to horticultural crops—the specific area familiar to both of us. Broadly speaking, this entails the range of crops covered by the abstracting journal *Horticultural Abstracts*. The decision is not entirely arbitrary or for convenience only. Horticultural crops have many attributes that make them unique and yet diverse (Table I). The entire range of disease-causing organisms and many animal pests affect horticultural crops. All plant parts are attacked, and the whole gamut of growers' attitudes is covered. These features place in sharp focus the questions as to why and when to model.

The chapter is developed by firstly considering the modelling background: what are models, what types of models are there, what are the objectives of models and what is the 'state of the art' in practical implementation? Secondly, some reasons why some models have found little application are presented: gaps between research and practice, attitudes of growers and the technological constraints are discussed. The discussion is supported by reference to a questionnaire survey of members of the East Malling Research Association (EMRA), an association of growers, corporate bodies and other interested parties in fruit research. Thirdly, the development of a conceptual

Table I. Characteristics of various types of horticultural crop.[a]

	Plantation	Fruit	Vegetable	Flower	Ornamental	Protected
Intensive management	+	++	+	++	++	+++
Single plants important	++	++	—	—	+++	—
Perennial	+++	++	—	—	+++	—
Clonal propagation	++	+++	—	—	+++	—
High value/ha	+	+++	+	++	+++	+++
Quality produce	+	+++	++	+++	+++	++
Marketing support	++	+++	+	++	+++	+++

[a] Key: —, a characteristic of no relevance; +, ++ and +++, characteristics of lesser, intermediate and higher relevance.

disease management model for apple will be described and discussed as an illustration of how a practical system for on-farm use may develop: experience in a disease management trial at East Malling will be heavily drawn on. Finally, our conclusions, not entirely pessimistic, will be presented.

II. THE MODELLING BACKGROUND

A. Model Typology

There is now a substantial literature on mathematical models in plant pathology. Although models in plant pathology have been developed at all levels of biological integration, most have been concerned with interacting host and pathogen populations, and with weather, that is, with epidemiology. Even within the discipline of epidemiology, there has been a wide variety of models. The sole criterion we shall use for accepting some entity as a model is that it can in principle be written in mathematical or symbolic form.

There have been many debates on the differences between the so-called theoretical and descriptive models (Jeger, 1983a), the former based upon inner logic and consistency in their construction, the latter based upon particular data. Similar dichotomies between so-called explanatory and predictive models have been drawn which disregard the fact that if a model is an explanation of some natural phenomenon, then it necessarily predicts. In fact these dichotomies rapidly degenerate into arid academic exercises and are not particularly useful when it comes to implementation. Models ought only to be evaluated from the standpoint of their purpose. If a theoretical model purports to be an explanation then it must be subjected to an empirical test; if a model is developed as a tool in an experimental programme then a subjective evaluation of its role in the programme is what is important; if a descriptive model is made to forecast some natural phenomenon then the accuracy of the forecast and the range of conditions over which this accuracy holds is all that is important. This point is not meant to be laboured, but merely to indicate that models for implementation can only be judged on their utility and the extent to which they achieve what they are supposed to do. The type of model is immaterial, except, for example, where its complexity prevents its implementation. It is far too early to determine what type of model is most useful in practical disease management.

An analysis of disease and pest models developed in Western Europe before 1982, restricted to cereals, top fruits, hop and grapevine, has appeared (Jeger and Tamsett, 1983). The analysis suggested that very few models have been developed for direct use on the farm, even though, it was claimed the majority could be used for forecasting. In the remainder of this section we examine a number of disease models representative of a wider geographical area, and consider some possible reasons as to why models fail to be implemented in practical management. The differences between models and operational systems are emphasised, and the limitations on their use discussed.

B. Models in Plant Pathology—Case Studies

To repeat, the main concern of this chapter is with models that have been, or have the potential to be, implemented on the farm. No restriction on the type of model is made. The models considered (Fig. 1) are concerned with incidence–intensity relationships, forecasting events within the epidemic cycle, forecasting disease progress, estimating quality and yield losses; and economic models, where the latter include techniques for optimisation, risk assessment and decision-making. The case studies are largely restricted to models developed in the period from publication of Kranz (1964) to the end of December 1982. To avoid overlap with Chapters 7 and 8, the construction of models, especially in subsections II,B,2, 3 and 4, is not considered in any

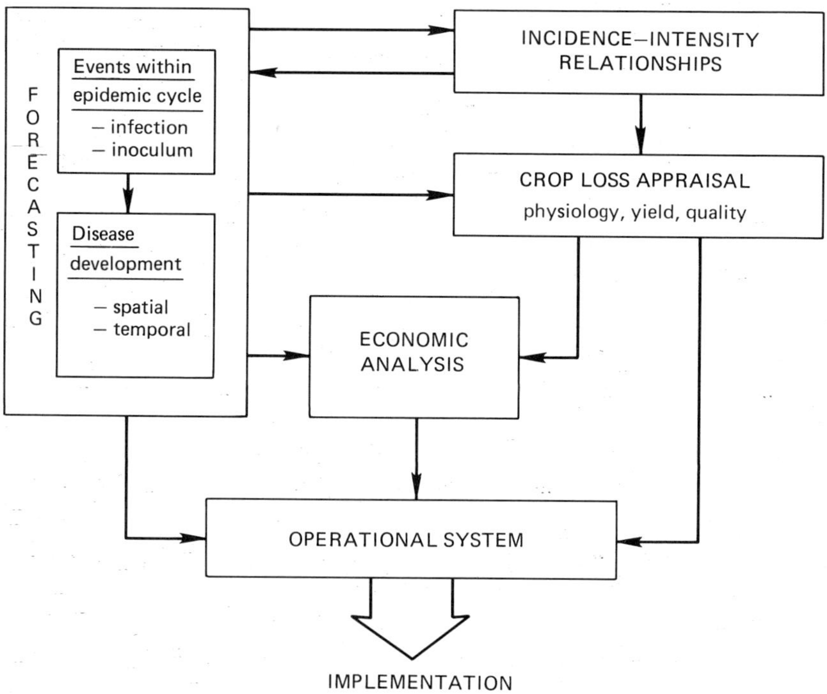

Fig. 1. The different types of models in plant pathology considered as case studies; their interrelationships and progress towards implementation.

detail, but rather the attributes of the models that may foster or hinder practical implementation.

1. Incidence–Intensity Relationships

Plant pathologists continually face the question of how to measure disease. Conventionally, two procedures may be discerned: determining a sampling unit such as leaf, shoot or tree, taking a sample, counting the number of units with lesions or other symptoms and expressing this number as a proportion (*incidence*); alternatively, the proportion of the unit area covered by disease lesions is estimated using a disease key, or the number of lesions per unit is counted (*intensity/severity*). The term 'intensity' is preferred (Butt and Royle, 1980) and is used throughout this chapter. The intensity of disease better reflects host damage, but is time consuming to measure and is prone to assessment error. Accordingly, if the relationship between intensity and incidence can be established independently and modelled, then incidence data can be used to estimate intensity.

The importance of the incidence–intensity relationship has long been recognised in crop loss studies (James, 1974) and should play a major role in disease management. The relationship is usually linear at low values of intensity with small changes of intensity resulting from large changes in incidence. Trends in incidence are a good indication of future damage to the host prior to this damage occurring. There is a period of time during which corrective chemical treatments can be applied. The supervised control scheme for apple powdery mildew developed at East Malling for growers depends, in part, upon first determining the incidence–intensity relationship (Butt and Barlow, 1979). Similar relationships play an important part in the EPIdemics, PREdiction and PREvention (EPIPRE) system of supervised disease management in winter cereals (Rijsdijk, 1982; Rabbinge and Rijsdijk, 1983).

The relationships between incidence and intensity may be developed from *a priori* assumptions or empirically. Often quite different models give good fits to their respective data but deviate significantly when compared (Jeger, 1984a). Major problems to be resolved before incidence–intensity relationships are implemented, as part of a disease assessment scheme, lie in the robustness of the models when different fungicide programmes or cultivars are compared (Seem and Gilpatrick, 1980), and in the extension of the principle to all plant parts attacked by pathogens (Jeger, 1984a).

2. Forecasting Events within the Epidemic Cycle

Disease epidemics can be considered as cycles of infection, colonisation, sporulation and dispersal. The components of the cycle can be modelled in their own right or be further partitioned *ad infinitum*. Models have been proposed for the 'subcomponents' of germination (Waggoner and Parlange, 1974), appressorial formation (Lapp and Skoropad, 1976), spore formation

(Pearson *et al.*, 1977) and hyphal growth and branching (Prosser and Trinci, 1979). We will restrict our attention to the macroscopic components discussed by Berger (1977) and Parlevliet (1979).

There has been some confusion over models that forecast disease and those that forecast infection *per se* (Krause and Massie, 1975). The former are based usually on weather that has occurred; the latter strictly demand that weather be forecast, and are an approach rarely followed (Gillespie and Sutton, 1979). In practice the function of most models is to warn of infection that has occurred. Eisensmith and Jones (1981a) devised an environmental favourability index for leaf spot of sour cherry from data collected in controlled environments. A threshold value of the index was obtained to delineate the minimum conditions for leaf infection. The index was used as a basis for timing curative sprays against the disease (Eisensmith and Jones, 1981b) and comparable results to scheduled spraying were obtained. Thomson *et al.* (1977) developed an equation to provide a warning of infection of pear flowers by the fireblight bacterium; it dependend on daily mean temperatures exceeding a threshold during a period of heavy bloom. The threshold decreased linearly during the spring. These two examples show several developments in this type of model; they may be developed from controlled environment rather than field observations, qualitative thresholds can be derived from quantitative models, and thresholds need not remain static. However, there have been few new examples of models forecasting components other than infection from weather, that add to the case studies reported by Butt and Royle (1974), or further information on how such models may be developed for implementation by growers.

The models based on weather inputs alone can be criticised in that no infection can occur without inoculum. There have been three main areas in which the amount of available inoculum has been used to forecast infection: soil-borne inoculum, spore trap data and the overwintering phase of foliar pathogens. There are two main problems associated with the monitoring of pathogen populations in the soil: the technical problems of propagule retrieval (Kobayashi *et al.*, 1981) and estimation of population density (Pfender *et al.*, 1981) are severe. The different interpretations of inoculum density–disease incidence relationships (Gilligan, 1983) make the implementation of models relating disease to soil-borne inoculum a dubious proposition. However, Adams (1981) found that the incidence of white rot of onions at harvest was quadratically related to the density of sclerotia in the soil. The equation was used for the evaluation of cultivars rather than for management of the disease. Again, the distinction between modelling disease development and infection *per se* should be recognised. In very few studies has inoculum densities been related to infection and then to disease incidence (Tomimatsu and Griffin, 1982).

Air-borne inoculum is easier to monitor, although problems concerning sampler efficiency and the representativeness of data remain. Various models have been proposed to relate infection or disease development to numbers of trapped spores (Berger, 1973). However, there is very little opportunity for growers to respond to such information, even supposing they have the ability to monitor spore production or receive the information from advisers or consultants. Another elaboration, discussed by Butt and Royle (1974), is to use weather and monitored disease to estimate the amount of air spora which is used in turn to forecast infection.

Many foliar pathogens, for example *Venturia inaequalis*, overwinter in the sexual stage on fallen leaves and discharge ascospores in spring when new foliage is most susceptible. Techniques for monitoring the development of the pseudothecia are technically demanding, time consuming and unlikely to be implemented on the farm or even by regionally based advisers. An alternative is to forecast the amount of spring inoculum from weather during the overwintering period. Equations describing ascospore maturation were developed by Massie and Szkolnik (1974) using a cumulative heat sum and rainfall from the time of leaf fall. Implicit in the model is the assumption that overall winter conditions are important. Conditions at leaf fall (Jeger and Butt, 1983b) and early spring (Gadoury and MacHardy, 1982), but not during the intervening period (James and Sutton, 1982), have been shown to be the determining factors. Each of these workers produced equations describing pseudothecial development or ascospore maturation but none, to our knowledge, has been implemented. Maturation is one aspect of overwintering: no equations have been developed that reliably forecast the abundance of pseudothecia from weather conditions (Jeger and Butt, 1983b), nor is there a clear relationship with assessed disease at leaf fall (Jeger, 1984b).

Data on inoculum and weather conditions should logically be used in any forecast of infection (Royle, 1973) or disease development (Murugesan *et al*., 1977) but, as with other models involving inoculum, the problems of monitoring severely restrict implementation. Inoculum and weather need not be the only variables used in descriptive models. Zelazny (1980) established relationships between the incidence of cadang–cadang disease of coconut palms and location variables such as elevation, abundance of other vegetation, palm density and soil conditions. Finally, many models originate without representation in a quantitative relationship. The Mills infection criteria, for example, were originally given as a table, although they can readily be expressed as equations (Mills and LaPlante, 1951). There are numerous examples in the literature where infection (Spotts, 1977), sporulation, viability and dispersal (Blaeser and Weltzien, 1979), incubation period (Karova, 1972) or disease development (Lim, 1972) are forecast without formal equations.

3. Forecasting Disease Development

An alternative approach is to consider entire disease progress curves and develop techniques for forecasting increases in time and space. The time span of the forecast is then an important feature for any subsequent implementation. For example, Kemp and Troup (1978) used accumulated degree-days and total 'bright sunshine hours' in April to forecast the incidence of non-persistent viruses of *Capsicum* at the end of August. Risk years could then be forecast in advance of planting—a most desirable feature in a model or forecasting scheme. Less lead time was provided by Murugesan *et al.* (1973), who forecast the incidence of phyllody disease of *Sesamum* on the ninetieth day after planting from mean minimum night temperature 30–60 days after planting. Shorter forecasts such as those provided by Sharma *et al.* (1979) properly belong with the examples in the previous sub-section.

Disease rarely occurs uniformly in growing crops. Distinct foci of disease occur, often with steep gradients (Gregory, 1968). There are few examples of gradients being used in disease management, although there are some examples in research (e.g. Fried *et al.*, 1979). The potential of gradient models is that decisions concerning spatial aspects of crop management can be made and implemented. Allen (1978) modelled the spread of banana bunchy top virus in established plantings with and without roguing of diseased plants. The model was successful in describing spread in plantations and has potential for determining quarantine measures to limit the spread of disease. Kable *et al.* (1980) studied the spread of peach powdery mildew from diseased rose thickets into peach orchards. The pattern of spread was adequately described by a gradient model, and the authors calculated necessary isolation distances for the siting of new orchards.

Models describing the temporal rather than the spatial increase of disease are more common due mainly to a longer history (Large, 1945; Van der Plank, 1963). Techniques for describing disease progress have proliferated and are most often used in *a posteriori* analysis rather than in real time. One of the few examples of temporal models of disease increase being used within a growing season is the simple exponential model, with a variable rate parameter, that forms the basis of the EPIPRE system (Rijsdijk, 1982). The monitoring and extrapolation of disease trends over a number of growing seasons has also been found useful (Ledger and Egan, 1979; Zelazny and Niven, 1980).

Techniques for analysing epidemic increase have usually been used for strategic purposes, such as comparing alternative strategies for control, rather than directly in management. Taylor *et al.* (1979) used an epidemic model to investigate the tolerance levels of bean seed to infection by *Pseudomonas phaseolicola*, and to compare the chemical treatment of seed and foliage: under United Kingdom conditions seed treatment would usually be of most benefit. Sall (1980) formulated a mathematical model, for grape powdery mildew including a host growth function, to examine the influence of seasonal weather and timing of initial infection on the eventual colonization of leaves and fruits. Although, conceivably, the model could be developed and implemented, its use has been limited to determining years most and least severe for the disease.

There is now a substantial literature on models based on systems concepts. Systems models have been constructed for diseases of horticultural crops (Waggoner and Horsfall, 1969; Kranz *et al.*, 1973; McCoy, 1977), but the rate of appearance of new models seems to have slowed down. Several reviews (Schrum, 1978; Kranz and Royle, 1978; Zadoks, 1979; Rabbinge and Carter, 1983) have discussed their possible but as yet undemonstrated roles in disease management. In our view there are major problems in implementing these models. First, the models soon become as complex as the system being modelled—which negates the very reason for modelling. Secondly, if the means to implement a model is available, then the model should be incorporated into general management packages concerned with the totality of agronomic practice (Gutierrez *et al.*, 1979; Kranz and Hau, 1980). These two considerations mitigate for the simplification of complex models before implementation (Rabbinge, 1983). There are, however, few examples of such simplification being achieved in practice. Shoemaker *et al.* (1979) incorporated a simplified disease model of olive knot bacterium into a management model for a range of pests of olives. Fry *et al.* (1983) gave the first demonstration that complex systems models in plant pathology can be used to define simplified forecasting schemes.

4. Forecasting Crop Losses

Much work has been concerned with developing valid methodologies for estimating loss (James, 1974). Virtually all loss models have been descriptive, with no in-built consideration of the yield physiology of diseased plants, and have been considered elsewhere (Madden *et al.*, 1981; Teng, this volume). Most studies and models have been concerned with cereal crops. We discuss crop loss with respect to horticultural crops to provide an alternative view of the subject.

A surprising number of models have been proposed for leguminous crops. Schneider *et al.* (1976) compared models for estimating yield losses of cow pea caused by *Cercospora* spp. Areas under disease progress curves were found to be correlated with yield losses, but the model did not give accurate estimates of losses at low levels of disease. A critical point model could only be used where disease progress was smoothly continuous. A hybrid model, incorporating features of the integrated curve and the critical point models, was developed and used for chemical control programmes. Kerr *et al.* (1978)

conducted surveys on the incidence of white mould disease on dry edible beans. All components of yield were reduced. Conditions favouring vigorous growth and high yields also favoured, however, white mould development. The net result was that low to moderate levels of disease were associated with higher yield, but high levels of disease severely reduced yield. A second-degree polynomial was necessary to describe this relationship. The inference that a certain level of disease is desirable makes the implementation of such models a delicate exercise in educating the grower. Sakai and Kono (1975) studied the losses caused by mosaic disease of Japanese radish: yield was linearly related to the time the disease was present in the crop. This parameter is often neglected in loss studies and would be considerably simpler to use in management.

Crops afflicted by single diseases are rare. Usually more than one disease is present and the assessment of losses due to a combination of different diseases is technically difficult. Effects on yield are rarely additive. Williams (1978) looked at the separate effects of chocolate spot and rust on broad beans. He proposed mathematical models relating severity to yield loss, but it is not clear whether those models included the full range of interactive effects of the diseases.

Perennial fruit and nut tree crops pose particularly severe problems for the forecasting of yield (Jeger, 1984a). Losses due to quality defects can be as important as yield reductions. Disease may affect yield and vegetative growth in future years rather than the present. There are complex relationships between components of yield and host growth. Very few studies of perennial crops have provided systematic rather than anecodotal data on crop losses. Butt *et al.* (1983b) have established relationships between the incidence of powdery mildew and various components of yield of 'Cox's Orange Pippin', the major commercial apple cultivar in the United Kingdom. The work has been of value in characterising both cost of control and yield responses in relation to disease, and in helping to determine thresholds for implementation in mildew management.

A major problem in crop loss assessment is the extrapolation of losses on experimental or commercial farms to larger regions. In very few areas are regions sufficiently homogeneous with respect to crop, cultivar, cultural practice, topography or weather to make possible the aggregation of losses using models. One possibility is to use aerial photography in conjunction with models of loss developed in field experiments (Wallen and Jackson, 1975; Basu *et al.*, 1978).

5. *Economic Models*

Most mathematical models developed in plant pathology have been descriptive linear regression models with few inputs, or complex systems models that attempt to mimic and synthesise all knowledge of the disease. There are very few examples of intermediate models that attempt to address specific questions with the simplest modelling tools available. As an example, objectives such as optimisation, risk assessment and decision analysis have been achieved in many technological disciplines using various mathematical techniques. Such techniques could usefully be used in the economic management of diseases. Economic aspects should not just be seen as an economic conversion of the output of a complex systems model. More direct approaches are required, using either simplified models (Mathys, 1974) or methods unfamiliar to most plant pathologists (Kahn, 1979; Shoemaker, 1981). Most importantly, economic aspects should be considered at the outset if implementation on the farm is to be achieved.

C. Operational Systems

1. Delivery and Support

A model developed in research is no guarantee that it will eventually be implemented by growers. The model has to be 'packaged' or placed in the context of an overall delivery and support system. Where advice based on model output is provided regionally by advisers or consultants, the means to communicate the advice to the grower must be available. Such systems, usually using telephone, radio or telex links, have been developed for apple scab in the United Kingdom (Adams and Seagar, 1977), France (Boue and Chaffurin, 1980; Geoffrion, 1980), the Republic of South Africa (Schwabe, 1980) and Canada (Coulombe and Jacob, 1981). Other operational systems have been developed for potato blight (Nutter and MacHardy, 1980), downy mildews of onion (De Weille, 1975) and *Cercospora* leaf spot of peanuts (Parvin *et al.*, 1974; Smith *et al.*, 1974).

In very few cases have models originating in research programmes been operationalised by their designers for implementation. Royle (1979) describes one example in the case of a forecasting scheme for hop downy mildew (Royle, 1973). If models are not computerised, then an easily digested means of presenting information to growers, such as manuals or charts (Ganry and Meyer, 1972; MacHardy, 1979), is needed as part of the operational system. The operational system should also include a range of diseases (Peries, 1973; Adams and Seagar, 1977). The most complete operational system devised so far appears to be in Michigan, USA, for pests and diseases of several crops (Brunner *et al.*, 1980).

2. Hardware

Many of the weather criteria necessary for infection can be expressed as mathematical functions. If weather data are recorded, stored and processed automatically, then microprocessor-based devices can be programmed to give disease warnings. Various devices have been developed, in particular for apple scab (Gendrier *et al.*, 1975; Richter and Haussermann, 1975; Tosic and Vasic, 1977; Grimm, 1977; Jones *et al.*, 1980; Gillespie and Kidd, 1980; de Siebenthal, 1981); several are operational and commercially available. The main drawback seems to be their inflexibility and dedicated, single-purpose use. Some of these objections have been met by the development of CDEM, a Crop Disease Environment Monitor (Sparks and Wass, 1983), in the United Kingdom. This device is designed for cereal diseases, apple scab, potato late blight and fireblight, and hence for the mixed farm enterprise. The implementation envisaged, however, is regional rather than on farms.

3. Limitations on the Uses of Models

There is now considerable evidence that descriptive weather-based models do not operate successfully outside the specific location where they were developed. Palti and Rotem (1973) show how cultural practices in arid regions may make areas more conducive to the development of peronosporaceous fungi and late blight than weather conditions and models based on weather would suggest. Billing (1979, 1980) has compared warning systems for fireblight from Europe and North America and found differences due to the patterns of weather in each region. A combined system for assessing risk was suggested, but it is not known whether this works as well as the individual models over all regions. Some workers have found simplified versions of models to be acceptable for different climates. Kable and MacKenzie (1980) found that the number of infection periods indicated by BLITECAST was considerably reduced in hot climates. Nutter and MacHardy (1980) used long-term records of weather to eliminate unnecessary features of BLITECAST in the northeast USA.

Models often fail where factors limiting or otherwise important in a region of implementation were not so limiting in the region where the model was constructed. Sutton *et al.* (1981), in a comprehensive evaluation of an apple scab ascospore maturity model developed in New York, USA, found that failure to forecast accurately was caused by an environmental variable important in North Carolina, USA, being unimportant in New York. The authors considered that model construction based on growth chamber studies, and an examination of a comprehensive range of all environmental variables, would avoid such model failures. Similar views are expressed by advocates of 'explanatory' systems models, but neither view seems to have been tested rigorously. For example, do systems models show any less susceptibility to failure than descriptive models when evaluated in a different environment or location? These important problems of model evaluation and, sometimes, calibration are delaying the transition of models developed in research to operational systems and hence to implementation on the farm.

III. PRACTICAL IMPLEMENTATION OF MODELS

The previous section has shown that very few models have been implemented or used directly in disease management. Of 58 models developed in Western Europe and analysed by Jeger and Tamsett (1983), only 3 were used directly on the farm. A total of 16 were implemented or handled by government advisory services such as the Meteorological Office or Agricultural Development and Advisory Service (ADAS) in the United Kingdom. Accordingly, the remainder of this chapter cannot be a report on practical implementation of models on the farm, but rather will discuss what, in our opinion, is required if implementation is to become a reality.

A. Why Implement Models On-farm?

The first question to be posed is why should models be implemented for management at the farm level? This is not a question directed at 'why model'? (We find the justification for modelling as an intellectual aim has been demonstrated many times in research, development and teaching.) In a sense models have always been implemented on-farm, although in most cases the criteria used are so trivial, obvious or else ill defined that rarely have they been considered models, for example, the determination of sowing dates. There are also examples of strategic models, resulting from research, being implemented in that general attitudes and goals of farmers are quite strongly influenced by the conclusions drawn from these models. The implementation of tactical, real-time models is a quite different proposition, but one to which we also take a positive attitude. Models enable or assist decision-making, models are a means of distilling complex information into a basis for action and models are a retrospective means of analysing the propriety of those actions. Recent developments in technology available to growers have made possible the automatic collection and processing of large amounts of information. It is not even necessary for the grower, self-consciously, to be aware that a mathematical model is being used—he will be operating a management system.

B. Why Are Models Not Being Implemented?

A wider question is, 'why are there information gaps between researchers and growers'? Norton (1982) considers instances in which the information needed by the grower is not available. The gaps can arise in different areas: *research*, in that the models are not available; *synthesis*, in that the models have not been simplified and 'packaged'; *dissemination*, in that there are no mechanisms for communicating to the grower; and *reception*, in that the model is neither clear not useable by the grower. It is dangerous to overgeneralise but our view is that the major gap with disease models is in the second area, synthesising. If models are to be implemented, then it is essential that packaging be done, whether by advisory services, by consultants, or by researchers in developing the models.

C. Role of the Grower

The customer of the management system, the grower, farm manager or advisor, should play an important role in the development and implementation of models. For this to occur it is essential that grower confidence in the system be obtained. Procedures for education and training must exist. Growers' attitudes ought to be surveyed so that models are constructed that are appropriate for the grower. The technical means for implementation and support of models in disease management systems must be available.

1. Grower Confidence

The grower must have confidence that management systems utilising models will be of benefit to him. This confidence must be obtained at all stages of model development from design through to the eventual packaging into the management system. In many cases, such as commodity-orientated agricultural research stations in the United Kingdom, the framework for developing such confidence exists in informal and formal links with grower organisations, advisory services and private consultants. The Members' Days of EMRA, the British Working Group on Integrated Crop Protection in Fruit and the biennial conferences of the Under 40s Group of Growers are all examples of contacts between researchers, growers, managers, government advisers and independent consultants concerned with fruit production in the United Kingdom. These meeetings and contacts must be used to maximum advantage by inspiring grower confidence in the development of disease management systems. Eventually, the grower needs confidence that the information obtained using the system will be of significant economic benefit (Webster, 1982).

2. Training and Education

Given a receptive grower, who is to train him in the use of disease management systems? These systems may involve diverse skills ranging from the operation and maintenance of delicate weather sensors and data loggers to direct interaction with an office computer. In our opinion the expertise and sophistication of growers of intensively managed crops, such as apples, make such a question almost an impertinence. Many growers already operate at a high level of management skill and the incorporation of disease models should present few problems, although, of course, the system must be designed with care. Many disease management schemes will require the monitoring of disease on the farm. This has always been considered a major obstacle. Researchers can help in overcoming this by quantifying incidence–intensity relationships, so that only incidence need be recorded, and by providing clear visual keys for the diagnosis of specific diseases, as is done, for example, with the EPIPRE system. A disease assessment method for apple powdery mildew was developed at East Malling for use by growers or advisers in orchards (Butt and Barlow, 1979), and has been well received and used. There is no reason to suppose that such monitoring cannot be linked with model implementation in a disease management system.

3. Attitudes of Growers to Disease Management

There seem to be two contrasting views on grower attitudes to disease management that depend upon the nature of the crop and the cropping system. On the one hand there are intensively grown, high input–high value crops such as apples; and on the other, low input crops with a low value per unit of land such as cereal crops. One view considers that only in the former can models or management systems be implemented, as only here does the grower stand to gain significantly, whereas there is little to gain in the latter crop. This may be called the *profit-maximising* view. The alternative view is that the first grower cannot be persuaded to implement models because he has so much to lose if the system fails, whereas the latter has much less to lose. This may be called the *risk-averse* view. In our opinion and limited experience the categorisation of growers into either of these extremes is not particularly useful. Apple growers combine both traits in their overall approaches to farm management.

4. What Types of Model?

The researcher can assist in the process of implementation by determining the problem that has to be faced on the farm and the kind of model appropriate for its solution. Do growers want quantitative or qualitative models? Do they want systems which give unambiguous recommendations, such as ‘spray or no spray’; or decision support, e.g. ‘if sprays are applied now there is a 95%

chance of yields being increased by 2%'? Very little modelling has started within the perspective of the grower rather than the researcher. Most importantly, the researcher has to recognise the scale of operation of the enterprise, the range of crops grown and the conflicting demands made. The researcher's favourite crop and disease may not be the grower's number one priority all of the time! This is why the flexibility of a microcomputer-based system with an ability to tune the system to local conditions and requirements seems more attractive to us than the dedicated single-purpose device. The system must also be operationally feasible. The grower must be able to act on the information obtained using the model. No matter how scientifically valid the information, it is useless in real-time management if the grower cannot make a decision and act within the time constraints imposed by the information. For example, curative spraying as a response to infection periods requires that appropriate chemicals be available and applied within a limited period of time.

5. Components of a Management System

Much of the hardware necessary for the implementation of simple, weather-based models is now available. Field data loggers and sensors are available that can be linked with microcomputers directly or indirectly by cassette or telecommunication interfaces. As noted earlier, dedicated 'black boxes', dealing with single diseases or disease combinations, are also available, although they will not be considered further. Many English apple growers do in fact collect weather data on the farm at present and the development of relatively inexpensive and automatic-weather stations should offer many advantages to the progressive growers. Similarly, many growers own or expect to own an office microcomputer within the next few years. These trends make possible a data base system of great power and flexibility when combined with other information systems such as Teletext or Viewdata (Houseman, 1982), or, for example, CASP (Walker, 1982). The main requirement for the implementation of these systems is software, the provision of which remains the great imponderable. The solutions to this problem will depend upon the particular characteristics, strengths and agricultural infrastructures of different regions and countries but will probably involve some combination of research, government advisory services and private enterprise. Researchers have already produced disease-management systems for use in research (Mogk, 1974; Eisensmith *et al.*, 1980). Government advisers and chemical companies have produced software for management 'games' to evaluate strategies of pest, disease and weed control (e.g. Jones, 1982). Educational games such as APPLESCAB have been devised. The EPIPRE system of supervised disease control in winter wheat is operational on a commercial basis in some countries in Western Europe (Rijsdijk, 1982; Rabbinge and Rijsdijk, 1983), although few details of its components have been made available.

The evaluation of management systems raises important questions for the implementation of models. It is not desirable that systems become operational without some independent evaluation of the system's working or utility. This evaluation must also apply to disease management systems provided by chemical companies. This may appear quite unrealistic but need not be, as chemical companies are quite prepared for independent evaluation of their fungicides. The dangers of full-scale implementation without independent evaluation are that sophisticated packaging (although clearly necessary) and the indirect benefits of implementation, such as enhanced grower awareness of disease problems through training and monitoring, may obscure inappropriate or invalid models that underly the system. System failure and bad experiences on the part of growers may lead to a premature rejection of the use of models in management systems and this may take years to rectify.

6. Questionnaire of EMRA Members

The management of apple mildew and scab was the topic at a Members' Day held at East Malling in 1981. Members were invited to complete a questionnaire concerned with their current practices, attitudes and perceptions of disease management, especially with regard to the implementation of models and disease management systems on the farm. We were encouraged by nearly 200 responses to the questionnaire, although these were probably biased in favour of the more progressive growers, and we make no claim to the rigour of the professional social scientist in designing the questionnaire or analysing the responses. In summary, we found that more than 60% of growers were prepared to move in the direction of decision-based management, most were favourable to on-farm monitoring of disease and weather (about 75% do in fact monitor weather on the farm at present), and about 40% expected to own an office microcomputer in the near future. The consensus view that emerged was in favour of a microprocessor-based device for monitoring and processing weather data, and for information to be displayed on the device, although some preferred the system to be linked to an office microcomputer. Warnings of conditions favourable to certain diseases and pests were needed but the system should have ancillary roles, such as giving information on water needs, frost risks and crop development. This combination of responses and the consensus that emerges indicates to us the potential for developing a disease management system for apple in the United Kingdom (Butt and Swait, 1984).

IV. A DISEASE MANAGEMENT SYSTEM FOR APPLE IN THE UNITED KINGDOM

Many of the points raised in the preceding section can be illustrated by looking specifically at apple growing in the United Kingdom; how a disease management system for research purposes, using models and other information tools, is being developed at East Malling and how this may provide a basis for practical on-farm implementation (Fig. 2).

1. Qualitative Analysis

There is a range of pests and disease affecting apple orchards. The major diseases of apple in the United Kingdom are scab and mildew, although canker (*Nectria galligena*) and collar rot (*Phytophthora cactorum*) can be devastating under certain conditions or cultural practices. A useful first step in analysing their effects is to construct an interaction damage matrix

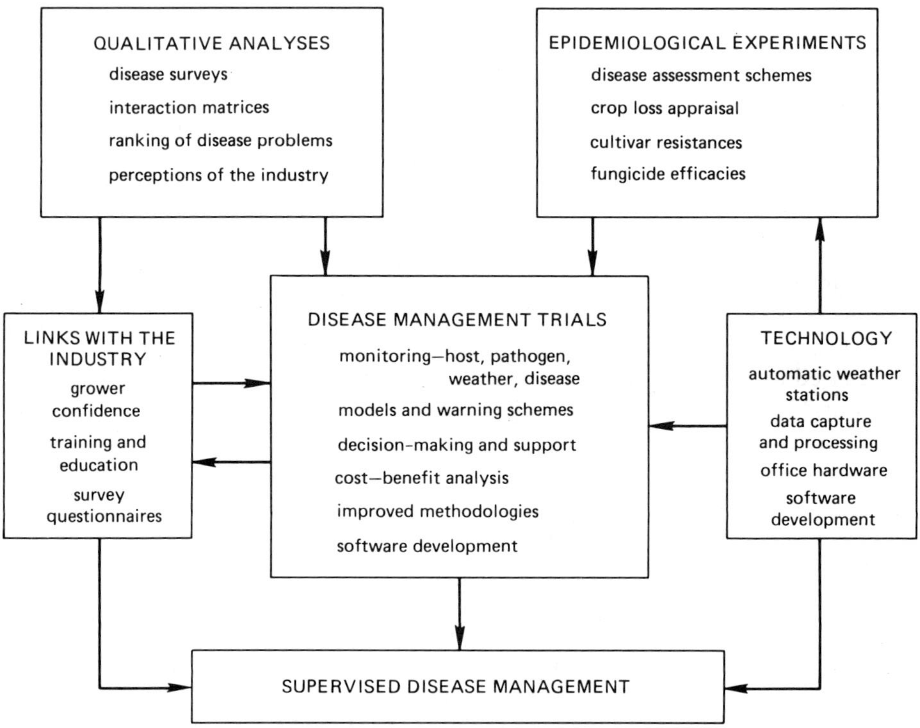

Fig. 2. Conceptual basis for the development of a disease management system for implementation on the farm. The example considered is apple growing in the United Kingdom, but the procedures described are quite general.

between host or orchard components and the various pests and diseases. An example of this procedure is provided by Norton (1983). Other examples of interaction matrices are between the pests and diseases: for example, many mildew fungicides are acaricidal and if used in orchards will negate the use of predator mites for controlling red spider mite. Another useful step is to solicit the views from the industry and obtain a rank order of pests and diseases as perceived by growers. The conclusions we draw from a qualitative analysis at East Malling are that disease and pest management should be based on good mildew control and the biological control of red spider mite. Other diseases and pests should be catered for within this overall strategy.

2. Experimentation

Experiments are required for the construction and development of models in disease management, and also to provide a data base to evaluate the models. The pilot disease management trial described in the next section started in 1979 and was improved by accompanying epidemiological experiments and the inclusion of more quantitative elements as data became available. The main experimental inputs to the disease management trial were disease monitoring, crop loss–mildew relationships, partial resistances of apple cultivars to mildew, epidemiological modes of action of mildew fungicides, integration of mildew fungicides with cultivars, overwintering of scab pseudothecia and scab conidial production from wood pustules. These experimental studies have been described by Butt and Barlow (1979), Jeger (1981, 1983b), Butt *et al.* (1981, 1983b), Butt and Jeger (1982), Jeger *et al.* (1982), Jeger and Butt (1983a,b) and in East Malling Annual Reports from 1977 *et sequens.*

3. Rocks Farm Plot Disease Management Trial

The disease management trial was commenced in 1979 in a commercial apple orchard at Rocks Farm, East Malling. Details of the rationale of the trial are given by Butt and Jeger (1982). Briefly, the trial consisted of two replicated treatments, corresponding to supervised and scheduled spraying, and an unsprayed control area. We made no attempt to use complex models in this trial, although these exist in part for scab (Kranz *et al.*, 1973; MacHardy and Jeger, 1983). We were more concerned with developing a management system in which, initially, decisions were made partly on a subjective basis and, later, on a more objective and quantitative basis as information became available from the accompanying experiments. A considerable amount of monitoring of host, disease, pathogen and weather was done that led to decisions on when and how to spray.

The major results of the first 3 years of the trial are summarised by Butt *et al.* (1983a). In general these results confirm the feasibility and economic benefits of supervised control in which models play a small but important part. Some chemical treatments for powdery mildew were needed even in wet and cool summer conditions. Overall fungicide costs for mildew were reduced by 30% and a total of nine sprays were saved in the supervised compared to routine plots. Tactical sprays of antisporulant fungicides at short (1–4 day) intervals proved an effective means of reversing an increasing trend in mildew incidence. Curative sprays for scab, based on Mills infection criteria, were less effective and more expensive than routinely applied protectant sprays. The need for scab assessments to be sufficiently sensitive to detect primary infections in the spring and the presence of lesions on wood was clearly shown. Yield in the routine plots exceeded that in the supervised plots in 2 years, but this was possibly due to a beneficial effect on fruit set of one fungicide used only in the routine plots.

Of more importance in the context of this chapter are several important and unresolved questions concerning the methodology of disease management trials. To what extent are such trials valid when the treatments imposed on a 'classical' experimental design are not physical dosages or rates of fungicide, or different cultivars, but rather approaches (e.g. supervised vs. scheduled spraying) or even attitudes (e.g. risk-aware vs. profit-maximising) to disease management? The task in the remaining period of the Rocks Farm trial is to simplify and synthesise the essential features of the disease management system and to make these available to growers in a cost-effective form.

V. CONCLUSIONS

We have discussed the features of different types of crops and their diseases. In many cases horticultural, especially perennial, crops with quality produce have unique features which make them suitable for evaluating the feasibility of implementing models and disease management schemes. We have concentrated most of our attention on horticultural crops. We have considered the range of models and modelling techniques that have been used by plant pathologists, and concluded that as yet few have made any direct impact on practical disease management. Many models, however, have the potential to make such an impact, but are limited more by a lack of synthesis or 'packaging' into management systems and by insufficient attention to the growers' viewpoint, rather than by inherent deficiencies in the models. This reflects on both the descriptive and the complex systems models. We do not specifically favour either type of model; both have a place in management systems at different stages of development. The final outcome, in many cases, will be a hybrid of these two categories of model.

A model implemented on the farm will be one component part of a disease management system for obtaining, processing, summarising and analysing data and other information. The end result is that something more than *ad hoc* decisions can be made, and action taken. At this stage of implementation the use of the term model can conveniently be forgotten; the emphasis will be on the total management scheme and will be concerned with objectives such as risk analysis, optimisation, design, decision making and forecasting. We are convinced that models in this guise will become more prevalent, and that growers are quite capable of using them. The main problems to be overcome lie in the production of software to enable the implementation of a management system, and the danger that sophisticated software may overimpress and obscure the need to evaluate properly the models underlying the system. Disenchantment or cynicism among growers is to be avoided at all costs. Experience at Rocks Farm, East Malling, gives us some confidence as to the role of models as one part of a disease management system, but we do not wish to minimise the many problems to be solved and constraints to be overcome before their role becomes practical rather than potential.

REFERENCES

Adams, P. B. (1981). *Phytopathology* **71**, 1178–1181.

Adams, R. J. and Seagar, J. M. (1977). *Meteorol. Mag.* **106**, 112–116.

Allen, R. N. (1978). *Aust. J. Agric. Res.* **29**, 1223–1233.

Basu, P. K., Jackson, H. R. and Wallen, V. R. (1978). *Can. J. Plant Sci.* **58**, 159–164.

Berger, R. D. (1973). *Phytopathology* **63**, 930–933.

Berger, R. D. (1977). *Annu. Rev. Phytopathol.* **15**, 165–183.

Billing, E. (1979). *Bull. OEPP* **9**, 45–51.

Billing, E. (1980). *Ann. Appl. Biol.* **95**, 365–377.

Blaeser, M. and Weltzien, H. C. (1979). *Z. Pflanzenkr. Pflanzenschutz* **86**, 489–498.

Boue, H. and Chaffurin, C. (1980). *Bull. OEPP* **10**, 223–226.

Brunner, J. F., Croft, B. A., Welch, S. M., Dover, M. J. and Bird, G. W. (1980). *Bull. OEPP* **10**, 259–268.

Butt, D. J. and Barlow, G. P. (1979). *Proc. Br. Crop Prot. Conf. 1979* **1**, 77–86.

Butt, D. J. and Jeger, M. J. (1982). *In* 'Decision Making in the Practice of Crop Protection' (R. B. Austin, ed.), pp. 167–190. British Crop Prot. Council, Croydon.

Butt, D. J. and Royle, D. J. (1974). *In* 'Epidemics of Plant Disease: Mathematical Analysis and Modeling' (J. Kranz, ed.), pp. 78–114. Springer-Verlag, Berlin and New York.

Butt, D. J. and Royle, D. J. (1980). *In* 'Comparative Epidemiology. A Tool for Better Disease Management' (J. Palti and J. Kranz, eds.), pp. 29–45. Pudoc Wageningen.

Butt, D. J. and Swait, A. A. J. (1984). *Proc. Br. Crop Prot. Conf. 1984* **2**, 655–663.

Butt, D. J., Jeger, M. J., Swait, A. A. J. and Souter, R. D. (1981). *Grower* **96**, 4–5.

Butt, D. J., Jeger, M. J. and Swait, A. A. J. (1983a). *Proc. Int. Congr. Plant Prot. 10th, 1983*, p. 1005.

Butt, D. J., Martin, K. J. and Swait, A. A. J. (1983b). *Proc. Int. Congr. Plant Prot. 10th, 1983*, p. 118.

Coulombe, L. J. and Jacob, A. (1981). *Phytoprotection* **62**, 44–52.

de Siebenthal, J. (1981). *Rev. Suisse Vitic. Arboric. Hortic.* **13**, 179–181.

De Weille, G. A. (1975). 'An approach to the possibilities of forecasting downy mildew infection in onion crops.' *Meded. Verh., K. Ned. Meteorol. Inst.* **97**.

Eisensmith, S. P. and Jones, A. L. (1981a). *Phytopathology* **71**, 728–732.

Eisensmith, S. P. and Jones, A. L. (1981b). *Plant Dis.* **65**, 955–958.

Eisensmith, S. P., Loria, R., Olsen, B. D. and Jones, A. L. (1980). *Plant Dis.* **64**, 646–651.

Fried, P. M., MacKenzie, D. R. and Nelson, R. R. (1979). *Phytopathol. Z.* **95**, 140–150.

Fry, W. E., Apple, A. E. and Bruhn, J. (1983). *Phytopathology* **73**, 1054–1059.

Gadoury, D. M. and MacHardy, W. E. (1982). *Phytopathology* **72**, 901–904.

Ganry, J. and Meyer, J. P. (1972). *Fruits* **27**, 665–676.

Gendrier, J. P., Fort, G., Mouchart, A. and Reboulet, J. N. (1975). *Def. Veg.* **29**, 240–248.

Geoffrion, R. (1980). *Bull. OEPP* **10**, 109–118.

Gillespie, T. J. and Kidd, G. E. (1980). *Can. J. Plant Sci.* **60**, 213–219.

Gillespie, T. J. and Sutton, J. C. (1979). *Can. J. Plant Pathol.* **1**, 95–99.

Gilligan, C. A. (1983). *Annu. Rev. Phytopathol.* **21**, 45–64.

Gregory, P. H. (1968). *Annu. Rev. Phytopathol.* **6**, 189–212.

Grimm, R. (1977). *Schweiz. Z. Obst- Weinbau* **113**, 88–92.

Gutierrez, A. P., Wang, Y. and Jones, R. E. (1979). *Bull. OEPP* **9**, 133–148.

Houseman, C. I. (1982). *In* 'Decision Making in the Practice of Crop Protection' (R. B. Austin, ed.), pp. 207–208. British Crop Prot. Council, Croydon.

James, J. R. and Sutton, T. B. (1982). *Phytopathology* **72**, 1073–1080.

James, W. C. (1974). *Annu. Rev. Phytopathol.* **12**, 27–48.

Jeger, M. J. (1981). *Ann. Appl. Biol.* **99**, 43–51.

Jeger, M. J. (1983a). *Bull. Int. Org. Biol. Control: West Palaearctic Reg. Sect. New Ser.* **VI**/2, 3–17.

Jeger, M. J. (1983b). *Proc. 7th Under 40's Top Fruit Conf., 9–11 Jan., 1983*, Long Ashton, England, pp. 58–64.

Jeger, M. J. (1984a). *In* 'Plant Diseases: Infection, Damage and Loss' (R. K. S. Wood and G. J. Jellis, eds.), pp. 225–235. Blackwell, Oxford.

Jeger, M. J. (1984b). *Trans. Br. Mycol. Soc.* **83**, 495–500.

Jeger, M. J. and Butt, D. J. (1983a). *Bull. Int. Org. Biol. Control: West Palaearctic Reg. Sect. New Ser.* **VI**/4, 111–122.

Jeger, M. J. and Butt, D. J. (1983b). *Ann. Appl. Biol.* **103**, 201–218.

Jeger, M. J. and Tamsett, J. (1983). *Bull. Int. Org. Biol. Control: West Palaearctic Reg. Sect. New Ser.* **VI**/2, 57–76.

Jeger, M. J., Swait. A. A. J. and Butt, D. J. (1982). *Ann. Appl. Biol.* **100**, 91–98.

Jones, A. L., Lillevik, S. L., Fisher, P. D. and Stebbins, T. C. (1980). *Plant Dis.* **64**, 69–72.

Jones, M. J. (1982). *In* 'Decision Making in the Practice of Crop Protection' (R. B. Austin, ed.), p. 208. British Crop Prot. Council, Croydon.

Kable, P. F. and MacKenzie, D. R. (1980). *Plant Dis.* **64**, 165–167.

Kable, P. F., Fried, P. M. and MacKenzie, D. R. (1980). *Phytopathology* **70**, 601–604.

Kahn, R. P. (1979). *Bull. OEPP* **9**, 119–130.

Karova, V. (1972). *Zashch. Rast. (Moscow)* **20**, 26–27.

Kemp, W. G. and Troup, P. A. (1978). *Can. J. Plant Sci.* **58**, 1025–1028.

Kerr, E. D., Steadman, J. R. and Nelson, L. A. (1978). *Crop Sci.* **18**, 275–279.

Kobayashi, K., Tanaka, F. and Kondo, N. (1981). *Nippon Shokubutsu Byori Gakkaiho* **47**, 29–34.

Kranz, J. (ed.) (1974). 'Epidemics of Plant Diseases: Mathematical Analysis and Modeling.' Springer-Verlag, Berlin and New York.

Kranz, J. and Hau, B. (1980). *Annu. Rev. Phytopathol.* **18**, 67–83.

Kranz, J. and Royle, D. J. (1978). *In* 'Plant Disease Epidemiology' (P. R. Scott and A. Bainbridge, eds.), pp. 111–120. Blackwell, Oxford.

Kranz, J., Mogk, M. and Stumpf, A. (1973). *Z. Pflanzenkr. Pflanzenschutz* **80**, 181–187.

Krause, R. A. and Massie, L. B. (1975). *Annu. Rev. Phytopathol.* **13**, 31–47.
Lapp, M. S. and Skoropad, W. P. (1976). *Can. J. Bot.* **54**, 2239-2242.
Large, E. C. (1945). *Ann. Appl. Biol.* **32**, 319-329.
Ledger, P. E., and Egan, B. T. (1979). *Sugarcane Pathol. News* **22**, 29–31.
Lim, T. M. (1972). *Proc. Rubber Res. Inst. Malays. Plant. Conf.*, pp. 169–179.
McCoy, R. E. (1977). *Proc. Fla. State Hortic. Soc.* **89**, 296–299.
MacHardy, W. E. (1979). *Plant Dis. Rep.* **63**, 199–204.
MacHardy, W. E. and Jeger, M. J. (1983). *Prot. Ecol.* **5**, 103–125.
Madden, L. V., Pennypacker, S. P., Antle, C. E. and Kingsolver, C. H. (1981). *Phytopathology* **71**, 685–689.
Massie, L. B. and Szkolnik, M. (1974). *Proc. Am. Phytopathol. Soc.* **1**, 140.
Mathys, G. (1974). *Bull. OEPP* **4**, 417–428.
Mills, W. D. and LaPlante, A. A. (1951). *N.Y. Agric. Exp. Stn. (Ithaca) Ext. Bull.* **711**, 18–22.
Mogk, M. (1974). *In* 'Epidemics of Plant Diseases: Mathematical Analyses and Modeling' (J. Kranz, ed.), pp. 55–77. Springer-Verlag, Berlin and New York.
Murugesan, S., Ramakrishnan, C., Kandaswamy, T. K. and Murugesan, M. (1973). *Madras Agric. J.* **60**, 492–495.
Murugesan, S., Chelliah, S. and Murugesan, M. (1977). *Madras Agric. J.* **64**, 22–28.
Norton, G. A. (1982). *In* 'Decision Making in the Practice of Crop Protection' (R. B. Austin, ed.), pp. 3–12. British Crop Prot. Council, Croydon.
Norton, G. A. (1983). *Bull. Int. Org. Biol. Control: West Palaearctic Reg. Sect. New Ser.* **VI**/2, 46–56.
Nutter, F. W. and MacHardy, W. E. (1980). *Plant Dis.* **64**, 1103–1105.
Palti, J. and Rotem, J. (1973). *Phytoparasitica* **1**, 119–126.
Parlevliet, J. E. (1979). *Annu. Rev. Phytopathol.* **17**, 203–222.
Parvin, D. W., Smith, D. H. and Crosby, F. L. (1974). *Phytopathology* **64**, 385–388.
Pearson, R. C., Aldwinckle, H. S. and Seem, R. C. (1977). *Can. J. Bot.* **55**, 2832–2837.
Peries, O. S. (1973). *Q. J. Rubber Res. Inst. Sri Lanka* **50**, 208–217.
Pfender, W. F., Rouse, D. I. and Hagedorn, D. J. (1981). *Phytopathology* **71**, 1169–1172.
Prosser, J. I. and Trinci, A. P. J. (1979). *J. Gen. Microbiol.* **111**, 153–164.
Rabbinge, R. (1983). *Bull Int. Org. Biol. Control: West Palaearctic Reg. Sect. New Ser.* **VI**/2, 31–45.
Rabbinge, R. and Carter N. (1983). *Bull. Int. Org. Biol. Control: West Palaearctic Reg. Sect. New Ser.* **VI**/2, 18–30.
Rabbinge, R. and Rijsdijk, F. (1983). *Bull. OEPP.* **13**, 297–305.
Richter, J. and Haussermann, R. (1975). *Anz. Schaedlingsk. Pflanzenschutz Umweltschutz* **48**, 107–109.
Rijsdijk, F. (1982). *In* 'Decision Making in the Practice of Crop Protection' (R. B. Austin, ed.), pp. 65–76. British Crop Prot. Council, Croydon.
Royle, D. J. (1973). *Ann. Appl. Biol.* **73**, 19–30.
Royle, D. J. (1979). *Wye Coll. Dep. Hop Res. Annu. Rep. 1978*, 49–56.
Sakai, Y. and Kono, T. (1975). *Hiroshima Kenritsu Nogyo Shikenjo Hokoku* **36**, 73–76.
Sall, M. A. (1980). *Phytopathology* **70**, 338–342.
Schneider, R. W., Williams, R. J. and Sinclair, J. B. (1976). *Phytopathology* **66**, 384–388.
Schrum, R. D. (1978). *In* 'Plant Disease: an Advanced Treatise' (J. G. Horsfall and E. B. Cowling, eds.), vol. 2: How Disease Develops in Populations, pp. 223–238. Academic Press, New York.
Schwabe, W. F. S. (1980). *Phytophylactica* **12**, 219–222.
Seem, R. C. and Gilpatrick, J. D. (1980). *Phytopathology* **70**, 851–854.
Sharma, S. L., Chowfla, S. C. and Sohi, H. S. (1979). *Indian Phytopathol.* **31**, 424–429.
Shoemaker, C. A. (1981). *IEEE Trans. Autom. Control* **25**, 1125–1132.
Shoemaker, C. A., Huffaker, C. B., and Kennett, C. E. (1979). *Environ. Entomol.* **8**, 182–189.

Smith, D. H., Crosby, F. L. and Ethredge, W. J. (1974). *Plant Dis. Rep.* **58**, 666–668.
Sparks, W. R. and Wass, S. N. (1983). *Bull. OEPP* **13**, 27–32.
Spotts, R. A. (1977). *Phytopathology* **67**, 1378–1381.
Sutton, T. B., James, J. R. and Nardacci, J. F. (1981). *Phytopathology* **71**, 1030–1032.
Taylor, J. D., Phelps, K. and Dudley, C. L. (1979). *Ann. Appl. Biol.* **93**, 167–172.
Thomson, S. V., Schroth, M. N., Moller, W. J. and Reil, W. (1977). *Calif. Agric.* **31**, 12–14.
Tomimatsu, G. S. and Griffin, G. J. (1982). *Phytopathology* **72**, 511–517.
Tosic, M. and Vasic, V. (1977). *Zast. Bilja* **28**, 167–172.
Van der Plank, J. E. (1963). 'Plant Disease: Epidemics and Control.' Academic Press, New York.
Waggoner, P. E. and Horsfall, J. G. (1969). *Bull. Conn. Agric. Exp. Stn. New Haven* **729**.
Waggoner, P. E. and Parlange, J. Y. (1974). *Phytopathology* **64**, 605–610.
Walker, J. (1982). *In* 'Decision Making in the Practice of Crop Protection' (R. B. Austin, ed.), p. 212. British Crop Prot. Council, Croydon.
Wallen, V. R. and Jackson, H. R. (1975). *Phytopathology* **65**, 942–948.
Webster, J. A. (1982). *In* 'Decision Making in the Practice of Crop Protection' (R. B. Austin, ed.), pp. 33–42. British Crop Prot. Council, Croydon.
Williams, P. F. (1978). *Proc. Aust. Plant Pathol. Soc. Workshop, Lincoln Coll., N. Z., August 1977*, 21-1–21-6.
Zadoks, J. C. (1979). *Bull. OEPP* **9**, 227–234.
Zelazny, B. (1980). *Phytopathology* **70**, 700–703.
Zelazny, B., and Niven, B. S. (1980). *Plant Dis.* **64**, 841–842.

10

MODELLING TO A PURPOSE

J. C. Zadoks* and R. Rabbinge†

**Laboratory of Phytopathology, Agricultural University, Wageningen, The Netherlands*

†Department of Theoretical Production Ecology, Agricultural University, Wageningen, The Netherlands

I. INTRODUCTION

This last chapter tries to grope with the future of mathematical modelling of crop disease. We have not attempted to review the literature objectively. Some literature reviews and textbooks have appeared during the last few years on the use of simulation and models in plant protection (Jeger and Rabbinge, 1983; Ruesink, 1976; de Wit and Goudriaan, 1978; Kranz and Hau, 1980; Jeffers, 1978). These and the foregoing chapters give an introduction to the methods and background of modelling and simulation in crop protection. The present essay renders the personal views of the authors on the future of modelling as applied to crop protection. Although these views may be biased, the authors feel that they serve to combine the dual objectives of scientific pursuit and public service.

II. MODELS IN CROP PROTECTION

A. Models

An inherent feature of biological science is the conceptualization of complex systems into levels of organization from lower levels such as molecules, organelles and cells to higher levels such as communities, populations and ecosystems (Loomis *et al.*, 1979; Zadoks and Schein, 1979). In this hierarchy, each level exhibits a characteristic behaviour that results from integration of sublevel processes under influences from the external environment.

In crop protection research the mechanistic basis for such a behaviour is being explored by reductionistic techniques, by seeking to isolate each sub-level process from the influences of higher levels and from competing elements at the same level. A mechanism of limited extent, such as disease progress, is identified, and its behaviour explained in terms of relations of sub-level mechanisms such as latent and infectious periods and environment. These relations, limited in number, are quantified. Quantitative integration of those mechanisms into an explanatory model of system behaviour, however, remains a task for the crop protectionist who has to deal with the whole crop instead of the composing elements.

Mathematical models are used as a method to effect such integrations. This approach has been favoured by the absence of other effective methodologies, by the emerging formalism of systems analysis and by the availability of computers. Two broad categories of models may be distinguished: same-level descriptive models and multi-level explanatory models. A wide range of descriptive models exists. Multivariate regression models, for example, are used for yield loss assessment under variable conditions (Large, 1954; Teng *et al.*, 1980). Most of these models are static, i.e. they do not involve a concept of time. Variables are integrated seasonal totals of yield loss, disease intensity and/or abiotic factors such as temperature or rainfall. Sophistication is improved by introducing some concept of time based, for example, on the calculation of developmental rates of crop and disease as a function of temperature and by increasing the numbers and precision of environmental parameters such as humidity, wind speed, radiation. There are time-dependent models too. In epidemiology the logistic equation,

$$\mathrm{d}x/\mathrm{d}t = r \cdot x \cdot (1 - x)$$

is widely used as an example of a descriptive population model. In the foregoing chapters this equation has been applied as a first introduction to population dynamics in crop protection.

The explanatory approach, emphasised above, is considerably more sophisticated. It employs dynamic models of the system hierarchy in an effort to provide explanation and prediction of integrated behaviour from detailed descriptive knowledge at a lower hierarchical level (de Wit, 1968; Loomis *et al.*, 1979). These explanatory models interpret the epidemic on the basis of detailed knowledge of structural elements at the lower hierarchical level. In epidemiological models of diseases, latent period (p), infectious period (i) and corrected basic infection rate (R_c) are determined in laboratory studies under well-defined conditions. Explanatory models bridge the gap between these detailed studies and field performance of the disease. Correspondence between model calculations (simulation) and field observations, using the independent abiotic variables as forcing variables, increases our insight and helps in designing new experiments in order to test model performance and to quantify input/output relationships.

Thus, construction of the model and the study of its behaviour in comparison with the performance of the real system result in an explanation of the system's behaviour. This quantitative explanation is of a completely different nature from the teleological explanation often found in biology. The teleological approach, which explains behaviour from the viewpoint of its supposed function or purpose, can, however, have great heuristic value, and can stimulate the search for relations applicable in dynamic simulation.

B. The Modelling Process

Model building should begin with a clear formulation of objectives concerning the use of the completed model. Biological systems are so complex that their models always represent a simplification and abstraction of the real system. The objectives determine the limits of the system and the degree of simplification. The inter-connection of variables and processes defining the system is facilitated and visualised by the use of relational diagrams. These schematic representations of the model define the functional relationships which are quantified on the basis of information from the literature or appropriate experiments. Parametrisation is the word often used for this building step. Computer models or other algorithms are the next step in the modelling process. Verification, testing whether or not the computer program in fact operates on input data in the intended way, is a further step. Next, the model is validated by comparing model predictions with independent observation. Different validation procedures are used (Rabbinge and Carter, 1983) to evaluate model behaviour in comparison with the behaviour of the real system. In most cases model behaviour and experimental results of field tests are compared visually, and this is normally sufficiently accurate. However, statistical tests using the χ^2-test or other methods to test the correspondence between measured data and model calculations are also used.

A validated model can be used in sensitivity analysis to test the relative importance of different input relations and to pin-point gaps in our knowledge. Sensitivity analysis involves changing the values of rates, variables and initial conditions in the model to determine their contribution to the behaviour of the model. After sensitivity analysis, simplification is usually possible. Simplification leads to summary models which are used for decision making and other managerial purposes. The above description of the modelling process is idealised. In practice, many steps are iterative, that is to say, forward and backward steps are made continuously. In many cases a summary model forms the starting point of a comprehensive explanatory model.

C. Objectives of Modelling

Explanation is the main objective of simulation modelling. The design of explanatory models forces one to conceptualise all relevant relationships, even if they have not yet been studied experimentally. A misfit at testing time may suggest that one or more relations are missing. In such cases, simulation is an aid in gap finding. This heuristic effect is one of the secondary aims of simulation. Another secondary objective arises from the experience that many experimentally determined relationships give too much information. Often simulation results would have been nearly as accurate with, for example, half the amount of input information. If so, simulation can give guidance in efficiency of experimental work. The redundancy of information can apply to the number of measurements per item as well as to the number of items measured (e.g. temperature and relative humidity and dewpoint).

These objectives, one primary and two secondary, are meaningful only when simulation and experimentation go hand in hand and feed each other. This *conditio sine qua non* is, unfortunately, often overlooked. The experimentation has to be made at the explanatory level as well as the level to be explained, to feed (relationships and parameters), verify and validate the model.

Apart from the scientific uses of simulation characterised by the word 'understanding', the agricultural and technical applications must be considered, summarised in the word 'action'. Summary models and decision rules based on simulation studies may be developed to guide growers and help them make decisions on chemical treatments. (Not treating is a decision, too.) In doing so, the dynamic simulation model becomes part of a supervised control system. The authors' interest focuses on this type of application.

D. Examples of Epidemiological Models

Assume a classical rust development model based on the equations of Van der Plank (1963; Zadoks, 1971), considering the host as a constant. Disease severity is expressed as the quotient, x, of diseased leaf area over total leaf area. Reinfection is governed by an empirical value called the daily multiplication factor, R_c. In the flow diagram of the simulation model the material flowing through a sequence of states is leaf area. Imagine that the empirical R_c is now to be replaced by reinfection simulated by means of a spore dispersal model, based on environmental physics (Rijsdijk and Rappoldt, 1980). In a dispersal model, the material flowing through a set of states could be spores.

Introduction of a spore dispersal model into the population growth model may cause difficulties because different units are used in each, namely, leaf area and spores. The 'coupling' of the lesion flow and the spore flow models can be realized through two 'couplers', the spore liberation per unit area of leaf and the lesion size per spore deposited, both specified per unit time (time step) and per leaf layer. Leaf layer comes in because infection of old leaves is not relevant, whereas young leaves do not yet support sporulation. In Fig. 1, the coupling is indicated by dotted lines, which 'translate' one material into another, and the couplers are symbolised by squares, in which the coupling dimensions are entered. The two flows, thus coupled, form an information cycle within an epidemiological model. At every time step, the program goes through this entire cycle. When crop structure is introduced the situation

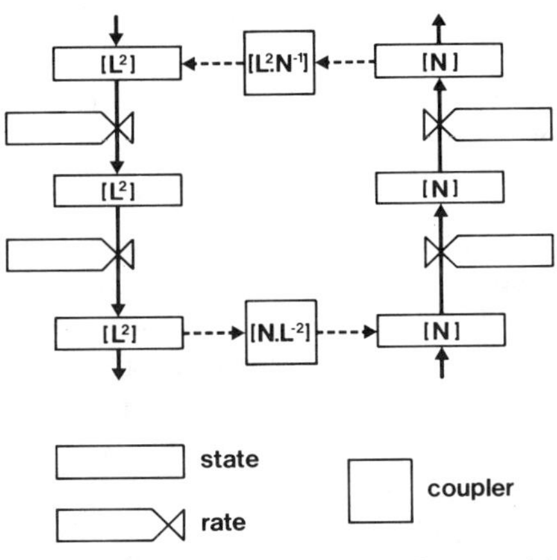

Fig. 1. Diagram of two coupled flows (heavy lines), one of area (e.g. leaf area, L^2) and one of numbers (e.g. spore numbers) coupled by two couplers, one translating leaf area into numbers (e.g. spores produced per cm^2) and one translating numbers into leaf area (e.g. lesion size in cm^2 per successful spore). In a similar way, two complex simulation models can be coupled, e.g. one of crop growth and one of spore dispersal.

becomes more complex. Leaf layers of different age differ in infectibility, latent period, infectious period and sporulation rate and this should then be introduced into the model (Rabbinge and Coster, 1984).

III. THE HOST CROP

A. Coupling of Models

To be realistic, epidemic models should consider the host as a variable (Zadoks and Schein, 1979). Growth and development of the host crop have to be simulated. Crop simulators exist in more (de Wit *et al.*, 1978) or less (Penning de Vries *et al.*, 1982) elaborate form, usually geared to the calculation of crop production under different conditions. In most epidemiological simulations, the crop is considered to operate at the optimum level, i.e. conditions are such that neither water nor other factors limit crop growth. This implicit assumption is in many cases invalid. de Wit (1982) distinguishes four levels of crop production, as follows.

Production level 1. Growth occurs in conditions with ample plant nutrients and soil water all the time. The growth rate is determined by the physiological characteristics of the crop and weather conditions.

Production level 2. Growth is limited by water at least part of the time, but when sufficient water is available the growth rate increases up to the maximum rate set by the weather.

Production level 3. Growth is limited by shortage of nitrogen at least part of the time and by water or weather conditions for the remainder of the growth period.

Production level 4. Growth is limited by the low availability of phosphorus, or by that of other minerals like potassium at least part of the time, and by nitrogen, water or weather for the remainder of the growth period.

For different production levels, different yields are expected and crop conditions are quite different. This affects the relation between host crop and disease. At each production level, pests and diseases do occur but the species may vary. The epidemiological characteristics and the damage relations of several pests and diseases are affected by crop condition (Cowan and van der Wal, 1975; Ayres and Zadoks, 1979). Thus diseases and pests of poor and rich crops exist (Zadoks, 1974; Rabbinge *et al.*, 1983).

B. Epidemiological Aspects

To simulate the epidemic of a foliar pathogen, specific data are needed such as leaf layers with their position along the stem, size, such as Leaf Layer Area

Index (LLAI), day of appearance and disappearance of leaves, inclination (angle), maturation (in relative age or in degree · days), and so on. All these items may affect infectibility expressed as lesions realised per germinated spore and sporulation capacity expressed as spores produced per lesion or per unit area. Specific information on leaves as required here is usually not explicitly provided by crop production models, but it can often be derived from them with moderate effort (van Keulen and Seligman, 1985). Once available, the crop model and the epidemiological model can be coupled.

Unfortunately, coupling of a comprehensive crop model to an epidemiological model is often a nuisance, because the two model types operate at different hierarchical levels with very different time steps. Such coupling is technically very difficult and usually it is not necessary to go to such pains. In many cases it is more profitable to proceed in steps. The comprehensive crop model is first reduced to a summary model. The summary crop model is used as a part of a combination model to calculate only those leaf data needed in the epidemics section of the combination model. As time steps of integration for simulation of crop data and for simulation of the epidemic may be different, combination models have to deal with so-called stiff equations (Goudriaan, 1973, 1977). To circumvent these problems different methods have been developed. The main program is run with relatively large time steps, whereas the sub-model describing the process with the small time coefficient is executed several times during each time step.

Combination models of crop growth and crop diseases are already producing realistic results and pave the way for the development of epidemiological summary models for use in crop management (Rijsdijk, 1982; Rabbinge and Rijsdijk, 1983). The stepwise approach described above results in useful summary models and reduces computer time and programming difficulties.

C. Injury–Damage Relationships

A further and more refined scientific aim is simulation of the effect of a disease on crop production. Seldom is there a straightforward relation between injury, assessed by eye, and damage, measured as reduction of marketable produce. If so, a feedback of pathogen on host by calculating destroyed leaf area is no longer sufficient. The pathogen affects the host in various ways, poisoning host leaves, promoting senescence or stimulating respiration. The pathogen may change the source–sink relations within plants and therewith influence the flow of assimilates, also into those sinks which later may become sources (Gaunt *et al.*, 1983). Thus, a feed-forward process comes into being.

Another problem is that lesions, even when growing under identical conditions, need not have equal effects (Zadoks and Schein, 1979). Complications arise when different production levels are considered. Cereal aphids on wheat, for example, cause much more yield loss per unit of aphid pressure at a high yield level than at a low yield level. The reason for this difference lies with the causes of yield loss (Rabbinge and Coster, 1984). Apparently, new objectives must be formulated to pursue our refined modelling: (1) the physiological effects of disease in response to production level need to be studied in detail, and (2) the coupling between host and pathogen models needs to be made at the elementary, physiological level. Such studies are in an incipient stage only.

D. Simulating the Farmer's Situation

Models as indicated above do not adequately represent the farmer's situation, because the farmer, at least in northwest Europe, always faces more than one harmful agent (Zadoks and Schein, 1979). It seems possible to use one summary crop model as a basis for several pest and disease models, at least when these are considered one by one, but this has rarely been done. The exception is probably cotton. Detailed models of the crop are combined with commensurately detailed models of insect pests (Gutierrez *et al.*, 1976; Peacock, 1980).

In the farmer's situation, however, diseases and pests do not develop independently. They influence each other in various ways; examples for wheat are

Puccinia recondita and *Septoria nodorum*	(van der Wal *et al.*, 1970; van der Wal and Cowan, 1974)
Pseudocercosporella herpotrichoides and *Puccinia striiformis*	(van der Wal and Zadoks, 1971)
Erysiphe graminis and *Griphosphaeria nivalis*	(Forrer *et al.*, 1982)
Lema melanopa and *G. nivalis*	(Forrer *et al.*, 1982)

The mutual influences cannot usually be coupled by way of leaf area infected, as they refer not only to the plant parts immediately affected but also to plant parts indirectly affected by way of a translocatable agent (*P. recondita* and *S. nodorum*) or of a nutritional feed-forward. For good modelling, the intrinsic route has to be followed at the physiological level.

Such multiagent–host models as suggested here do not yet exist. They may be constructed in the future. The constraint is not computer hardware or software, but the basic knowledge on input relations. To get this information, much detailed and painstaking research is needed. Most of this research is yet to be done. A possible application of the multiagent model will be discussed in the following section.

IV. PRACTICAL APPLICATIONS OF SIMULATION MODELS

A. 'Managerial Models'

It is not so easy to apply simulation models in practice: the most obvious use is in prediction, strategic or tactical (Shoemaker, 1981). Strategic models refer to long-term approaches, often have a general validity and indicate the consequences of certain pest and disease control policies. The choice of variety, production level, regional characteristics, etc., may come into a strategic model. Tactical models refer to the short run and the individual field. They react to the stimuli of the moment and need to be applied frequently during the growing season. The pest and disease manager, in many cases the farmer, wants to make a decision on whether to treat or not and he likes to decide on the basis of calculated alternatives. Strategic models offer him different policies and the consequences thereof. Tactical models are tailored to a specific option and the ensuing specific needs of their users. Unfortunately very few, if any, simulation models are 'managerial models' that predict not only the effect of fungicide applications on damage but also the financial results of managerial actions. The study of the epidemiological mode of action of fungicides has been neglected, even though it has been demonstrated that the results can be used in simulation models (Zadoks, 1982). The economics of fungicide applications have rarely been simulated, if at all.

The question which now arises is not whether it is possible to make such elaborate simulators but whether it is sensible to do so. The argument against elaborate managerial models is threefold: (1) They will be large and therefore expensive to run as an on-line per-field service, (2) they will provide a superfluous precision for some aspects which cannot be validated, and (3) they will lead to a reduced attention for other effects sometimes more important than pests or pathogens, such as considerations of profitability. Models for managerial use should contain a bare minimum of detail, with as few variables as possible. The first argument could be overcome by the construction of a 'summary model'. The second and third arguments are more fundamental. The problem is to prove that additional state variables are necessary and do not reduce attention from other important aspects. This proof can only be provided by an iterative working method combining theoretical and experimental work. The accuracy of field trials in crop protection is, however, rarely better than 5% difference in yield. If this error is expressed in terms of money its value often exceeds the added value of one treatment.

An example will clarify the point. Around 1980, Dutch wheat yields were on average about 7 metric tonnes per hectare. The accuracy of field trials is at best some 5%, or about 350 kg ha^{-1}. The costs of a treatment are also about 350 kg ha^{-1}. (By the way, 350 kg ha^{-1} is less than 2 days' dry matter production.) As the accuracy of a field trial money-wise is in the order of one treatment, the simulation can be verified up to an accuracy of $\pm\frac{1}{2}$ treatment at best. When gross yield is used to check the accuracy of the simulation, as is the normal case, the attainable accuracy lies within the limits corresponding with a money value of ± 1 treatment.

Greater precision in testing the accuracy of the simulation can be obtained by using other field data assessed during the growing season. The choice of the data depends on the objective of the study. For managerial purposes, the 'treatment', whatever its nature, is the entity to decide upon. Its added value is here considered as a 'unit of currency'. In management, precision better than $\pm\frac{1}{2}$ treatment is a waste of effort.

B. Decision Making

For managerial purposes superfluous precision is a nuisance rather than an advantage. This and other reasons may lead to simplification, from comprehensive model over summary model to decision models. The latter can take the form of decision trees or networks with a limited number of yes–no alternatives, ending in at most three options: treat, do not treat, or wait and see. The large number of time steps in the comprehensive model is reduced to the relatively few moments that a manager can profitably decide to do something or not to do it. The original simulation model should help to develop these decision criteria.

Decision models can be arrived at along other paths than simulation. An example of a decision model in disease forecasting, based on the recognition of critical periods, is the elaborate, computerized, field-specific BLITECAST against *Phytophthora infestans* on potatoes (Krause *et al.*, 1975). PHYTPROG, also computerized but not field specific, uses a form of field monitoring (Burckhardt and Freitag, 1969). Both examples contain decision criteria but they are not simulators. Spray recommendations using simulators seem feasible only when the simulator of disease is initialised for the specific conditions of the field for which the recommendation is wanted.

A decision model such as EPIPRE (Rabbinge and Rijsdijk, 1983; Rijsdijk, 1982; Zadoks, 1981, 1983), originating in part from elaborate simulation models, still reflects its simulation origin. Simplification was obtained by checking only whether or not a damage threshold would be surpassed, instead of calculating the whole course of the epidemic. In one version, it had some 25 decision points. In a newer version, further simplification is obtained by calculating the added values for a number of options and selecting the highest one.

C. Optimisation

Decision systems such as BLITECAST and EPIPRE attempt to optimise disease control but they do not yet utilise mathematical optimisation techniques. During the last decade optimisation techniques have been introduced into crop protection at several places (Shoemaker, 1981). Textbooks indicate how and where to use these techniques (Jeffers, 1978).

Classic optimisation theory tended to study problems of industry with few but large enterprises. In agriculture, the theory should adjust to a large number of small enterprises, with different perceptions of risk. Only recently, stochastic elements have been introduced in these models in order to compute chances for optimal solutions. Given a certain combination of constraints and a set of environmental factors, and given the variation in these environmental factors, the model turns out optimal solutions with their chance densities. Stochasticity enables the user to decide upon the degree of risk (in money) he is prepared to accept within a chosen control strategy and to see the costs involved. In optimisation, decisions can be made for the whole growing season at once, or stepwise by deciding for each optimisation period. In both cases the consequences of a certain decision have to be calculated. Dynamic simulation models can be used for part of the necessary computations (Onstad *et al*., 1983). Simulation and optimisation are complementary tools in pest and disease management!

V. LONG-TERM PERSPECTIVES

A. Crop Management Models

Farmers are not interested in one-disease-only recommendations. Pest and disease control is just one of the many crop husbandry actions to take. Many farmers consider this part of crop management as relatively unimportant, even though crop protection involves about 15% of their time and expenditure (Zadoks, 1980). They have to deal with a crop situation in which several choices have to be made. The varieties they choose, the crop husbandry practices they apply and the weather they experience affect crop growth and diseases and pests differentially. In addition, modern pesticides have plural effects; for example, triadimefon gives good protection of wheat against *P. striiformis* and *E. graminis* and fair protection against *P. recondita* and *S. tritici*. Others, such as sulphur, protect apple leaves against mildew (*Podosphaera leucotricha*) but kill beneficial predatory mites, and so on. 'Multi-agent + crop' managerial models in whatever form are a necessity indeed for farmers' acceptability and thus for further progress. Such models should become available in the mid-1980s.

Several other crop management decisions affect harmful agents, such as weed control, fertilizer applications and irrigation. EPIPRE was designed to recommend omission of the last N top dressing of wheat when *P. striiformis* had already attained a certain critical severity. The future will see comprehensive crop management models of which the crop protection aspects will be just one portion. In the Netherlands, work is in progress on aspects of timing and amount of nitrogen fertilisers and herbicides for wheat. Gradually, a complete crop management system will be developed. The main objective of this management system is maximisation of financial returns, but as societal aims change, pesticide minimisation may become another objective affecting management policies. EPIPRE already contains environmental thresholds in rudimentary form (Zadoks, 1983). Irrigation has already been incorporated for cotton, including a link with crop protection (Gutierrez *et al.*, 1976). Other economic or convenience options, such as an early and short picking period versus a more regular and prolonged picking period, are in operation now (Peacock, 1980), and these options have their specific crop protection implications. Before 1990, we will probably see supervised crop management systems in cotton and wheat, and possibly in a few other crops too, with higher farmer acceptability. Here, acceptability means three things: (1) the farmer is able to understand the background and the reason for the recommendation; (2) he is willing to accept a calculated risk; and (3) he is willing to pay a fee for a recommendation which may save him money. When the objectives of farmers or of society at large change, new optimal policies can be determined by running the dynamic optimisation program once more.

B. Centralised and Decentralised Systems

At the time of writing, 1983, most computerised management systems are centralised systems. This means that a central computer, far away from the target field, does the necessary calculations. Communication between farmer and computer is by mail (EPIPRE), by phone (BLITECAST) or by visiting consultant carrying a microcomputer that can interact with the main-frame computer (APPLESCAB). Faster interaction can be organised through VIDITEL systems, connecting home television screen and central computer by the public telephone network; such systems also permit two-way interaction between farmer and central computer.

Decentralised systems with on-the-farm and in-the-field devices will be implemented before 1990. Precursors are the various gadgets helping the grower to decide whether or not to treat against *Venturia inaequalis*. More sophisticated are the small $1000 black boxes called BLITECASTER, placed in the potato crop (MacKenzie and Schimmelpfenning, 1978). They monitor the microclimate of the crop and they inform the farmer according to the

BLITECAST program when sprayings are needed, thus obviating the necessity to communicate with the central computer. The next step is the installation of on-the-farm microcomputers which run not only farm administration programs, but also crop protection or even total crop management programs. Of course, these microcomputers can be connected with sensors placed in the field. Thus, the farmer will expand his diagnostic equipment and feel more independent. He will have his answers faster than by way of the central computer. For EPIPRE, a microcomputer adaptation for on-the-farm use has been developed. One disadvantage of decentralisation is the absence of communication between the farmer and the research organisation which is responsible for the recommendations and advisory system. A more serious disadvantage lies in the difficulty of updating a program rapidly in response to changes in the pathogen such as the appearance of a new race or a fungicide-resistant strain. A combination of centralisation and decentralisation seems feasible and might reduce the disadvantages of both.

VI. INTEGRATIONISTIC SCIENCE

The developments described in this chapter will stimulate a great deal of holistic research, that is research placing small elements of knowledge in their proper perspective or as 'proportional' parts of complex but operational bodies of knowledge. Maybe such research will help somewhat to reduce the redundancy of present scientific information, and at the same time help to identify and fill gaps in our knowledge. Such research will also help to bridge the present gap between 'understanding' and 'action'. If so, the lost feeling of 'scientific proportion', as John Grainger (1979) put it so nicely, will be restored. The integrationistic approach, as contrasting to much of the current reductionistic research activities, is one of the great contributions of dynamic simulation to modern science.

REFERENCES

Ayres, P. G. and Zadoks, J. C. (1979). *Physiol. Plant Pathol.* **14**, 347–361.

Burckhardt, H. and Freitag, E. (1969). *Der Kartoffelbau* **20**, 176–180.

Cowan, M. C. and van der Wal, A. F. (1975). *Neth. J. Plant Pathol.* **81**, 49–57.

de Wit, C. T. (1968). *Landbouwk. Tijdschr.* **80**, 215–218.

de Wit, C. T. (1982). *In* 'La Productivité des Pâturages Sahéliens'. pp. 20–35. (F. W. T. Penning de Vries, and M. A. Djitèye eds.), Pudoc, Wageningen.

de Wit, C. T. and Goudriaan, J. (1978). 'Simulation of Ecological Processes'. *Simul. Monogr.* Pudoc, Wageningen.

de Wit, C. T., Goudriaan, J., van Keulen, H., van Laar, H. H., Louwerse, W., Penning de Vries, F. W. T., Rabbinge, R. and Sibma, L. (1978). 'Simulation of Assimilation, Respiration and Transpiration of Crops'. *Simul. Monogr.* Pudoc, Wageningen.

Forrer, H. R., Rijsdijk, F. H. and Zadoks, J. C. (1982). *Neth. J. Plant Pathol.* **88**, 123–125.

Gaunt, R. E., Lim, L. G. and Thompson, W. J. (1983). *FAO Plant prot. Bull.* **30**, 3–8.

Goudriaan, J. (1973). *Neth. J. Agric. Sci.* **21**, 269–281.

Goudriaan, J. (1977). 'Crop Micrometeorology: a Simulation Study'. *Simul. Monogr.* Pudoc, Wageningen.

Grainger, J. (1979). *Annu. Rev. Phytopathol.* **17**, 223–252.

Gutierrez, A. P., Falcon, L. A. and van den Bosch, R. (1976). *In* 'modelling for Pest Management, Concepts, Techniques and Application' (K. Tummala, D. L. Haynes, and B. A. Croft, eds), pp. 134–145. Michigan State Univ. Press, East Lansing.

Jeffers, J. N. R. (1978). 'An Introduction to Systems Analysis with Ecological Applications'. Arnold, London.

Jeger, M. J. and Rabbinge, R. (eds.) (1983). *Bull. IOBC/WPRS NS* 1983/VI/2.

Kranz, J. and Hau, B. (1980). *Annu. Rev. Phytopathol.* **18**, 67–83.

Krause, R. A., Massie, L. B. and Hyre, R. A. (1975). *Plant Dis. Rep.* **59**, 95–98.

Large, E. C. (1954). *Plant Pathol.* **3**, 128–129.

Loomis, R. S., Rabbinge, R. and Ng, E. (1979). *Annu. Rev. Plant Physiol.* **30**, 339–367.

MacKenzie, D. R. and Schimmelpfenning, H. G. (1978) *Am. Potato J.* **55**, 384–385 (abstr.).

Onstad, D. W., Rabbinge, R. and Rossing, W. (1983). *Proc. Br. Crop Prot. Symp. 1983* I, 191.

Peacock, W. J. (1980). *Agric. Gaz. N.S.W.* **91**, No. 4.

Penning de Vries, F. W. T. and van Laar, H. H. (eds.) (1982). 'Simulation of Plant Growth and Crop Production'. *Simul. Monogr.* Pudoc, Wageningen.

Rabbinge, R. and Carter, N. (1983). *Bull. IOBC/WPRS NS* 1983/VI/2, 18–30.

Rabbinge, R. and Coster, G. (1984'. *Proc. Aust. Appl. Entom. Res., 4th,* 163–169.

Rabbinge, R. and Rijsdijk, F. H. (1983). *Bull. OEPP.* **13**, 297–305.

Rabbinge, R., Sinke, J. and Mantel, W. P. (1983). *Meded. Fac. Landbouwwet. Rijksuniv. Gent* **48/4**, 1159–1168.

Rijsdijk, F. H. (1982). *Proc. Br. Crop. Prot. Symp. 1982,* pp. 65–76.

Rijsdijk, F. H. and Rappoldt, K. (1980). *Ber. Umweltbundesamt* **5/79**, 407–410.

Ruesink, W. G. (1976). *Annu. Rev. Entomol.* **21**, 27–44.

Shoemaker, C. A. (1981). *IEEE Trans. Autom. Control* **26**, 1125–1132.

Teng, P. S., Blackie, M. J. and Close, R. C. (1980). *Agric. Syst.* **5**, 85–103.

van der Plank, J. E. (1963). 'Plant Diseases: Epidemics and Control'. Academic Press, New York.

van der Wal, A. F. and Cowan, M. C. (1974). *Neth. J. Plant Pathol.* **80**, 192–214.

van der Wal, A. F. and Zadoks, J. C. (1971). *Proc. Int. Symp. Plant. Pathol., 2nd.*

van der Wal, A. F., Shearer, B. L. and Zadoks, J. C. (1970). *Neth. J. Plant Pathol.* **76**, 261–263.

von Keulen, H. and Seligman, N. G. (1985). In press.

Zadoks, J. C. (1971) *Phytopathology* **61**, 600–610.

Zadoks, J. C. (1974). *Phytopathology* **64**, 918–923.

Zadoks, J. C. (1980). *Landbouwk. Tijdschr.* **92**, 313–323.

Zadoks, J. C. (1981). *Bull OEPP* **11**, 365–369.

Zadoks, J. C. (1982). *In* 'Fungicide Resistance in Crop Protection' (J. Dekker and S. G. Georgopoulos, eds.) pp. 149–160. Pudoc, Wageningen.

Zadoks, J. C. (1983). *Ciba Found. Symp.* **97**, 116–125.

Zadoks, J. C. and Schein, R. D. (1979). 'Epidemiology and Plant Disease Management'. Oxford Univ. Press, New York.

Index

E

Q

R

S

T

U

V

W

Y